MODALITY AND MEANING

Studies in Linguistics and Philosophy

Volume 53

Managing Editors

GENNARO CHIERCHIA, *University of Milan*
PAULINE JACOBSON, *Brown University*
FRANCIS J. PELLETIER, *University of Alberta*

Editorial Board

JOHAN VAN BENTHEM, *University of Amsterdam*
GREGORY N. CARLSON, *University of Rochester*
DAVID DOWTY, *Ohio State University, Columbus*
GERALD GAZDAR, *University of Sussex, Brighton*
IRENE HEIM, *M.I.T., Cambridge*
EWAN KLEIN, *University of Edinburgh*
BILL LADUSAW, *University of California at Santa Cruz*
TERRENCE PARSONS, *University of California, Irvine*

The titles published in this series are listed at the end of this volume.

MODALITY AND MEANING

by

WILLIAM G. LYCAN
University of North Carolina at Chapel Hill

KLUWER ACADEMIC PUBLISHERS
DORDRECHT / BOSTON / LONDON

A C.I.P. Catalogue record for this book is available from the Library of Congress.

ISBN 0-7923-3006-4

Published by Kluwer Academic Publishers,
P.O. Box 17, 3300 AA Dordrecht, The Netherlands.

Kluwer Academic Publishers incorporates
the publishing programmes of
D. Reidel, Martinus Nijhoff, Dr W. Junk and MTP Press.

Sold and distributed in the U.S.A. and Canada
by Kluwer Academic Publishers,
101 Philip Drive, Norwell, MA 02061, U.S.A.

In all other countries, sold and distributed
by Kluwer Academic Publishers Group,
P.O. Box 322, 3300 AH Dordrecht, The Netherlands.

Printed on acid-free paper

All Rights Reserved
© 1994 Kluwer Academic Publishers
No part of the material protected by this copyright notice may be reproduced or
utilized in any form or by any means, electronic or mechanical,
including photocopying, recording or by any information storage and
retrieval system, without written permission from the copyright owner.

Printed in the Netherlands

In memoriam

Janet Grace Lycan, 1905-1991

TABLE OF CONTENTS

Preface   xiii

Acknowledgements   xiii

Sources and Abstracts   xv

PART I   MODALITY

Chapter 1. The Trouble with Possible Worlds.   3

    1. Meinong vs. the Forces of Decency   4
    2. The Theoretical Status of Nonexistents   6
    3. The Real Problem   10
    4. Continuing Hostilities   11
    5. Approaches to Possibility   14
    6. The Meinongian Quantifier as Nonobjectual   16

Chapter 2. Three Conceptions of Possible Worlds.   25

    1. A Thumbnail History   25
    2. Lewis' Reversal   27
    3. The Independence of the Indexicality Thesis 30
    4. Lashed to the Masts   34
    5. Lewis' Argument Against Impossibilia   39

Chapter 3. Ersatzing for Fun and Profit.   45

    1. Potential World-Surrogates   45
    2. Combinatorialism   47
    3. Problems for Combinatorialism   48

4.	Tractarian Combinatorialism	55
5.	Prognosis	58

Chapter 4. Against Concretism.   73

1.	Understanding, or Not Understanding, Relentless Meinongianism	73
2.	Some Disadvantages of Relentless Meinongianism	75
3.	Mature Lewis' Modal Realism	80
4.	Three Further Objections to Mature Lewis	83
5.	Lewis Against Us Ersatzers	87
6.	Summing Up	90

Chapter 5. Essences.   95

1.	The Case for Haecceitism	95
2.	A Visit with Quine	99

Chapter 6. Fiction and Essence.   109

1.	The Conservative Line	109
2.	Opening Defense of the Conservative Position	110
3.	McMichael's Issue	113
4.	The Modal Properties of Fictional Individuals	115
5.	McMichael's Argument	117
6.	More on Behalf of the Conservative Position	118
7.	Yes, Haecceities for Nonactuals	120
8.	Two Further Options	123
9.	Geach's Puzzle about Intentional Identity	124
10.	Edelberg's Asymmetrical Cases	127

TABLE OF CONTENTS ix

Chapter 7. The Paradox of Naming Resolved by a Kinder, Gentler Theory of Direct Reference. 135

  1. *The Paradox and the Possibilities* 136
  2. *Plantinga's Compromise* 139
  3. *The Plantinga Problem Generalized* 140
  4. *A Paratactic Approach to Attitude Contexts* 142
  5. *A Two-Scheme Theory* 145
  6. *Abbreviation and the Spot-Check Test* 149
  7. *Direct Reference* 150
  8. *Frege's Puzzle* 152
  9. *Empty names* 156

Chapter 8. Relative Modality 171

  1. *Modalities in Natural Language* 172
  2. *Immediate Philosophical Returns* 178
  3. *Moral Psychology* 185

PART II MEANING

Chapter 9. Semantic Competence, Funny Functors, and Truth-Conditions. 203

  1. *Logical Space* 204
  2. *Frege's Horizontal* 207
  3. *Substitutional Quantification* 211
  4. *What Is Semantical Muteness?* 217
  5. *A New Counterexample* 218
  6. *Grades of Understanding* 221
  7. *Indexicals* 223

Chapter 10. Logical Constants and the Glory of Truth-
    Conditional Semantics. 233

  1. *Davidson's Wonderful Project* 233
  2. *"Meaning Postulates"* 235
  3. *"Logical Constants"* 237
  4. *The Underlying Magnitude* 241
  5. *Meaning Postulates Reinstated* 243
  6. *The Damage* 244
  7. *Denouement* 245

Chapter 11. Propositions and Analyticity. 249

  1. *Quine Against Analyticity* 250
  2. *The First Unpromising Account* 251
  3. *The Second Unpromising Account* 254
  4. *Truth by Convention* 256

Chapter 12. Stipulative Definition and Logical Truth. 263

  1. *Stipulative Definition* 263
  2. *The Objections* 264
  3. *The Ann Arbor Defense* 273
  4. *And so what?* 279

Chapter 13. Analogy and Lexical Semantics. 283

  1. *Polysemy* 285
  2. *Ross' Apparatus* 288
  3. *Metaphor* 291
  4. *Two Criticisms* 294
  5. *Repercussions and Alleged Repercussions* 298

TABLE OF CONTENTS	xi
BIBLIOGRAPHY	305
NAME INDEX	327
SUBJECT INDEX	333

PREFACE

Like its immediate predecessor, *Judgement and Justification* (Lycan [1988]), this book is neither a straightforward collection of previously published essays nor an entirely new book. Much of the material it contains has appeared previously, here or there (mostly there); but the material has been reorganized, expanded and supplemented, and there are a number of new chapters and sections.

In Part I I present and defend a coherent metaphysics of modality, an "Actualist" theory of possible individuals and possible worlds. Part II takes up some issues in truth-conditional semantics, particularly lexical semantics, that were not addressed in my *Logical Form in Natural Language* (Lycan [1984]).

As always, my footnotes are too many and too long, but as always, the text may be read smoothly without them.

ACKNOWLEDGEMENTS

This book was completed during my tenure as a Fellow of the Center for Advanced Study in the Behavioral Sciences, in 1991-92. I thank the Center for its outrageously generous support, financial, clerical, and (multiply) otherwise. For additional funding I am indebted to the National Endowment for the Humanities (#RA-20037-88) and the Andrew W. Mellon Foundation.

In the Preface to his book *Inquiry* (1984), Robert Stalnaker records that his year at the Center "inspired in [his] son the ambition to be, when he grows up, a philosopher on leave." My then twelve-year-old daughter took a similarly rosy view, adding frequent references to my "being paid to stuff [my] face and play volleyball." Well, I did finish this book, and another one. (*She* should talk, anyway.)

ACKNOWLEDGEMENTS

I am especially grateful to Leslie Lindzey and Carol Baxter for their arduous and skillful work on the preparation of my camera-ready hardcopy. I caused them a great deal of trouble, which they bore with unflagging good cheer. Upon my return from the Center, I received further valuable editorial help from Andrew Mills, Wayne Riggs and Eric Barnes.

For discussion, criticism, and helpful comments on previous versions, I owe thanks to many people, who include David Armstrong, David Austin, Lynne Rudder Baker, Dorit Bar-On, John Bigelow, Steven Boër, Max Cresswell, Michael Devitt, Jay Garfield, Susan Hale, Michael Hand, Frank Jackson, Michael Jubien, Murray Kiteley, Robert Kraut, Saul Kripke, David Lewis, Barry Loewer, Penelope Mackie, Richard B. Miller, Charles Pigden, Alvin Plantinga, W.V. Quine, Phil Quinn, Jay Rosenberg, Tony Roy, David Sanford, George Schlesinger, George Schumm, Stu Shapiro, Keith Simmons, Bob Stalnaker, Richard Sylvan, Pavel Tichy, Jim Tomberlin, Peter van Inwagen, Takashi Yagisawa, Ed Zalta, the late Herbert Heidelberger, and an anonymous referee for Kluwer. I have tried to acknowledge more specific debts in footnotes.

SOURCES AND ABSTRACTS

Most of the chapters are based on previously published essays, but almost all have been substantially edited and revised, and some have been kaleidoscopically intermixed with others. Here are the original sources, and abstracts of the present versions.

>Chapter 1: Selected portions of "The Trouble with Possible Worlds," in M. Loux (ed.), *The Possible and the Actual* (Ithaca: Cornell University Press, 1979), pp. 274-316.
>
>The trouble with "nonexistent objects," "nonactual possible worlds" and the like is not ontological, but is the problem of giving clear and separate meanings to the actuality-restricted quantifier and the apparent merely possibilist quantifier that both occur in Meinong's slogan "*There are* things that do not *exist*." Various approaches to that problem are surveyed, and a case is made that the two quantifiers should continue to be taken at face value as objectual quantifiers.
>
>Chapter 2: A considerably expanded version of "Two--No, Three--Concepts of Possible Worlds," *Proceedings of the Aristotelian Society*, Vol. XCI Part 3 (1990-1991), pp. 215-227.
>
>Once it has been settled that quantification over mere possibilia is literal and objectual, one must choose between an Actualist reinterpretation of the possibilist quantifier and a more outrageous Concretist interpretation. Surprisingly, even within Concretism one must choose between a genuinely Meinongian view and David Lewis' unexpectedly different version.

But once one makes one's fundamental choice between Actualism, Meinongian Concretism and Lewis' view, most of the hot debates current in modal metaphysics are easily settled; thus they are not independently real issues.

Chapter 3: Selected portions of "The Trouble with Possible Worlds" (*loc. cit.*) plus a chunk of "Armstrong's New Combinatorialist Theory of Modality," forthcoming in K. Campbell, J. Bacon and L. Reinhardt (eds.), *Ontology, Causality, and Mind* (Cambridge: Cambridge University Press).

Various Ersatzist proposals are surveyed. Combinatorialism is taken as a case study and examined at some length, but is rejected in favor of using sets of structured propositions as Ersatz "worlds." There is an Appendix on D.M. Armstrong's fictionalist twist.

Chapter 4: Selected portions of "The Trouble with Possible Worlds," plus my review of David Lewis, *On the Plurality of Worlds* (*Journal of Philosophy*, Vol. LXXXV (1988), pp. 42-47), plus some new material.

Both forms of Concretism are criticized at length. The contrasting virtues of Ersatzing are touted.

Chapter 5: My half of "Actuality and Essence" (with Stewart Shapiro), *Midwest Studies in Philosophy*, Vol. XI (Minneapolis: University of Minnesota Press, 1986), pp. 343-377; plus some new material.

Within possible-worlds semantics, essentialism in some form is *obviously* true of actual entities. For actual individuals, Haecceitist essentialism in particular is defended against any "qualitative" essentialism. The chapter then revisits the 1960s' issue of the relation between possible worlds, quantified modal logic, and metaphysical essentialism, in the new light of Kripke's notion of a rigid designator.

Chapter 6: Greatly expanded from the Appendix to "Actuality and Essence," *loc. cit.*

Although a qualitative essentialism at first seems obviously true of nonactual entities, Haecceitism turns out to be a very live possibility for nonactuals also. A rich though fully Actualist ontology of fictional characters is sketched.

Chapter 7: The first half based on "The Paradox of Naming," in B.-K. Matilal and J.L. Shaw (eds.), *Analytical Philosophy in Comparative Perspective* (Dordrecht: D. Reidel, 1985), pp. 81-102, and "On Respecting Puzzles About Belief Ascription (Reply to Devitt)," *Pacific Philosophical Quarterly*, Vol. 71 (1990), pp. 182-188; the second half is new.

A semantic theory of proper names is presented, designed to fill the need for a middle way between Millianism and Russell's Description theory. The account is very close to current "Direct Reference" versions of Millianism, but departs from them on one crucial and very beneficial point. The chapter also defends a view of empty names that is consistent with the truth-theoretic program for semantics.

Chapter 8: Appears here for the first time.

No ordinary English sentence expresses an unrestricted alethic modality. All everyday modalities are relative to contextually determined sets of background assumptions. The syntactic behavior of modal auxiliaries in English reflects this semantic/pragmatic context-dependence. Once the relativity is recognized, a number of surprisingly disparate traditional philosophical issues are shown to dissolve or at least become tractable, especially issues in moral psychology: e.g., the compatibility of human freedom with causal determination; "'Ought' implies 'can'"; and the ancient question of "Why we should be moral."

Chapter 9: Amalgamates "Semantic Competence and Funny Functors," *Monist*, Vol. 62 (1979), pp. 209-222, and "Semantic Competence and Truth-Conditions," in E. LePore (ed.) *New Directions in Semantics* (London: Academic Press, 1986), pp. 143-155.
   Syntactically unusual but well-formed counterexamples show that having a truth-condition does not suffice for meaningfulness; hence even a "structured meaning" in Cresswell's sense (a truth-condition hyperintensionally structured by a syntactic breakdown of the sentence in question) does not determine the sentence's meaning. Reasons for this are explored.

Chapter 10: *Notre Dame Journal of Formal Logic*, Vol. 30 (1989), pp. 390-400.
   The distinction drawn in linguistic semantics between strictly "logical" implication and merely lexical implication, or between logical truths and "meaning postulates," is bogus. The difference is one of smooth degree. The point has some bad consequences for the Davidsonian semantic program: a pattern of semantic explanation made famous by Davidson's "The Logical Form of Action Sentences" is seen to be far less interesting than has been thought.

Chapter 11: The first half of "Definition in a Quinean World," in J.H. Fetzer, D. Shatz and G. Schlesinger (eds.), *Definitions and Definability: Philosophical Perspectives* (Dordrecht: Kluwer, 1991), pp. 111-131.

Philosophers have been convinced or (at least) cowed into decades of mutinous silence by Quine's famous attack on analyticity. Yet, oddly, most philosophers reject the nihilistic view of meaning that most directly supports the attack on analyticity: Quine's scorning of meanings or propositions and his doctrine of the indeterminacy of translation. This chapter begins to make a Quinean case against analyticity that depends neither on scorning propositions nor on the indeterminacy doctrine.

Chapter 12: Second half of "Definition in a Quinean World," plus two new sections.

The Quinean case against analyticity continues. The best candidate for analyticity--truth by explicitly stipulative definition--is impugned. A Quinean view of definition is presented. A recent argument of Paul Boghossian's forces a major concession, but the spirit of the Quinean view is preserved.

Chapter 13: Based on my Critical Study of James Ross' *Portraying Analogy*, *Linguistics and Philosophy*, Vol. 11 (1988), pp. 107-124.

Lexical Atomism is the view that sentence meanings are syntactically compounded from the literal lexical meanings of a finite, manageably-sized stock of morphemes, those lexical meanings being (synchronically) more or less fixed, as listed in a dictionary. This chapter attacks Lexical Atomism, on grounds of rejecting the "literal"/"figurative" distinction presupposed by it. Virtually all lexical meaning depends, by virtue of the analogy mechanisms that generate "figurative" coinage, on intrasentential environment.

Necessity makes me suffer constantly,
And custom makes it easy...

--The Duchess of Malfi

PART I

MODALITY

CHAPTER 1

THE TROUBLE WITH POSSIBLE WORLDS

In what sense or senses, if any, should we admit that "there are" possible but nonexistent beings or possible but nonactual worlds? Sources of motivation for some such admission are powerful and various. By positing nonexistent individuals, it seems, we may understand true negative existentials and accommodate the intentionality of certain mental entities. By positing nonactual worlds or states of affairs, we may achieve our familiar but still remarkable reduction of the alethic modalities to quantifiers,[1] formulate truth-conditional semantics for propositional attitudes and hosts of other troublesome constructions, display the otherwise mysterious connections between Fregean senses and linguistic meaning,[2] illuminate the pragmatics of counterfactuals and other conditionals, and provide a rigorous format for the theoretical study of decision making.[3] Even ordinary ways of speaking encourage us to reify nonexistent possibles at every turn.[4]

And yet many philosophers are uneasy about yielding to this encouragement; and many, despite all the foregoing benefits, openly scorn the idea of a thing or world that has the property of being nonactual. What is striking is that for most of the present century, the dispute between the friends and the foes of mere possibilia consisted largely of intuitive ventings, dogged repetition of slogans, mutual accusations of perverse or willful misunderstanding, bad jokes, and simple abuse. Only in the 1970s did philosophers begin trying thoughtfully and painstakingly to get to the bottom of the problem. In this chapter and the next, I shall correct some common misunderstandings of the issue, set it up in what I think is the neatest and most illuminating way, distinguish some fundamentally different approaches to the vindication of nonexistent possibles, attack several prominent recent instances of these approaches, survey further prospects, and point in the direction that seems to me most promising.

1. MEINONG VS. THE FORCES OF DECENCY

As I understand him, Meinong took it to be intuitively obvious or self-evident that there are nonexistent possibles and even nonexistent impossibles.[5] We refer to such things by means of names and descriptions; and they are the objects of thought, after all. Serious researchers doing science and philosophy concern themselves primarily with what is really true, of course; this is as it should be, but unfortunately it produces in them a bias toward the actual and a blindness toward the other sectors of our ontology.

Russell, and Quine some decades later, expressed distaste for this way of looking at things.[6] Russell faulted Meinong's "sense of reality" and accused him of "doing a disservice to thought." Quine called Meinong's ("Wyman's") universe "overpopulated," "unlovely," "rank," a "slum," and "a breeding ground for disorderly elements." Russell and Quine thus found mere possibilia repugnant and wondered how Meinong could stomach them.

Meinong would not have been impressed by these gestures of distaste. He would simply have repeated his observation that philosophers who find nonexistent possibles aesthetically offensive have unhealthily (if understandably) restricted their diets to the actual.

Without doubt, metaphysics has to do with everything that exists. However, the totality of what exists, including what has existed and what will exist, is infinitely small in comparison with the totality of the Objects of knowledge. This fact easily goes unnoticed, probably because the lively interest in reality which is part of our nature tends to favour that exaggeration which finds the non-real a mere nothing--or, more precisely which finds the non-real to be something for which science has no application at all or at least no application of any worth. [1904/1960, p. 79]

Besides, Meinong would remind his critics, the ordinary person speaks quite familiarly and often of possible things that do not exist; this is as natural as breathing and as palatable as (real) beer. Russell and Quine have mongered the mystery, not he, and the cause of their doing so is their forgetting how such talk proceeds when the nature of the actual in particular is not what is specifically at issue.

A bit more can be said, however, about why Quine and Russell find mere possibilia so uncongenial. For one thing, notice that when in "A Theory of Objects" (1904/1960) Meinong does try to spell out a vocabulary for talking about nonexistent possibles and impossibles, his discussion takes on a theoretical and moderately technical tone. Certainly an explanatory system is being envisioned and limned, though Meinong himself (I believe) regarded his work on Objects as purely descriptive;[7] and if Meinong's ontology is an explanatory system, it is the sort of thing that must be justified on grounds of elegance and coherence as well as by its explanatory power. Moreover, as soon as Meinong does begin to talk a bit more technically, his apparatus raises questions whose answers are not obvious and which call for ad hoc ramifications on his part. For example: (i) Quine notoriously demands identity and individuation-conditions for mere possibilia. When have we one possible man and when have we two? When have we 8,003,746? (ii) Meinong seems to assume that any (well-formed) superficial singular term refers either to an existent or to a nonexistent being, and he certainly believes that any nonexistent being has any property expressed by the matrix of any admissible description used to refer to it.[8] This prompted Russell (1905a) to ask whether *the existent round square* exists. Similarly, we might ask whether *the Object that has no Sosein* has a *Sosein*, and so on. (iii) The same assumption requires that the city that is five miles north of Columbus and five miles south of Cleveland is five miles north of Columbus and five miles south of Cleveland. Does this not in turn entail that Columbus is ten miles south of Cleveland and so falsify Meinong's theory? (iv) Meinong's Objects are indeterminate or *incomplete* in a well-known way: given virtually any nonexistent possible O there will be any number of properties P such that it is not a fact that O has P and not a fact that O lacks P. Now, how can there be, or even "be," a man who is tall but whose height does not fall into any specific range? (v) Meinong characteristically refers to his Objects by using *definite* descriptions, such as "the golden mountain." But on Meinong's own view there are many golden mountains, such as the one which has a beebleberry bush on top, the one which has no beebleberry bush on top, the one on which the Marines stage practice assaults, and the one on top of which Descartes wrote his *Meditations*. How can our phrase, "*the* golden mountain," then succeed in uniquely denoting a single Meinongian Object?

A partisan of possibilia can set about to answer questions like these easily enough. Any elaborate way of herding possible men etc. into full-fledged worlds will settle question (i) determinately (e.g. that of Lewis [1986]); Kripke's (1972) or Rescher's (1975) stipulative methods would work just as well. In answer to (ii) we might impose a type theory on Meinong's ontology that would rule out such troublesome "descriptions" as ill-formed.[9] Routley (1980, pp. 268ff.) responds to (iii) by enforcing a distinction between "entire" and "reduced" occurrences of relational predicates. (iv) may be handled by locating Objects within worlds, or by explicating their nature in such a way as to make their "incompleteness" familiar (as does Castañeda [1974, 1989]). (v) leaves us any number of feasible alternatives. So Meinong need not be at all impressed by the fact that Russell and Quine can raise some trick questions for him. And there is some tendency in the recent literature to believe that Quine's objection to possibilia, at least, is based entirely on these trick questions. Any friend of possibilia who thinks this and who has answers to the questions will of course conclude that Quine should have no further quarrel with nonexistents. But such a person would miss the thrust of Quine's opposition entirely, as I shall now explain. (The position I shall now ascribe to Quine never appears explicitly in his writings, but it falls trivially out of well-known views of his.)

2. THE THEORETICAL STATUS OF NONEXISTENTS

We have conceded that questions such as (i)-(v) are readily answerable by the theorist. But that is exactly the point: that a theorist is needed to answer them. Meinong's view has generated the questions without any help from Russell or Quine, and any answers to them will necessarily involve elaboration of a theoretical apparatus. Any such apparatus, along with the proclaimed nature of the possibilia it posits, will have to be justified on theoretical grounds and submit to evaluation of the sort to which any philosophical theory is subject.

One concern that we have about theories is that of parsimony. Most philosophers subscribe to Occam's principle or something like it; at least, few philosophers posit entities that they admit to be totally gratuitous for purposes of philosophical explanation. Now, consider Meinong's ontology. Quine accuses it of bloated unloveliness. But the problem here is even worse than Quine explicitly observes: If Meinong's ontology is bloated, it is bloated to the bursting point. In fact, it is bloated well *past* the bursting point; Meinong believes not only in all the things there could possibly be, but also in all the things there could not be. And the point is not just that Meinong has swallowed some indigestible entities. The problem is that it is now hard to retain any use for Occam's Razor at all. A Meinongian has *already* posited everything that could, or even could not, be; how, then, can any subsequent brandishing of the Razor be to the point? How can the Meinongian explain the continuing usefulness (some would say the indispensability) of parsimony principles in philosophy and in science?[10]

We know how Meinong would respond: he would accuse us again of aiming our tunnel vision only at the actual. Everyone agrees that we must not posit entities *as existing* if they are not needed for purposes of explanation; Occam's Razor certainly applies to existents. But, Meinong would point out, his remaining Objects are *non*existents, and so they are unscathed by Occam's Razor or Occam's Stomach Pump or whatever.[11]

The suggestion is that we should posit possibilia, but not posit them "as existing." Quine's reaction is classic:

Wyman, by the way, is one of those philosophers who have united in ruining the good old word 'exist.' Despite his espousal of unactualized possibles, he limits the word 'existence' to actuality--thus preserving an illusion of ontological agreement between himself and us who repudiate the rest of his bloated universe.... Wyman, in an ill-conceived effort to appear agreeable, genially grants us the nonexistence of Pegasus, and then, contrary to what *we* meant by nonexistence of Pegasus, insists that Pegasus *is*. Existence is one thing, he says, and subsistence is another. The only way I know of coping with this obfuscation of issues is to give Wyman the word 'exist'. I'll try not to use it again; I still have 'is'. So much for lexicography; let's get back to Wyman's ontology. [1948/1963, p. 3]

The important point here is not that Quine does not care which verbs we use to mark ontological distinctions. Meinong does not care which terms might be used to mark his ontological distinctions either; all he insists is that the distinctions are real. Quine may simply announce that he does not care about the distinctions and go on to repeat his charge of bloating. All Meinong can do in response is to repeat his answer to that charge: that we genuinely commit ourselves to, and thus need be parsimonious about, only the Objects that we claim to find in *this* world, and not those which are merely objects of thought--thought is free, after all, and talk is cheap.

Contrary to what is suggested by the quoted passage, I believe Quine's real point has nothing to do with ordinary language: It is that anyone who actually makes theoretical use of a Meinongian apparatus in an appropriately regimented theory of modal semantics *quantifies over* nonexistent possibles in his official canonical idiom. Quine notoriously does not care what we say in casual speech; but the moment a semanticist such as Kripke or Montague writes the backward E, the semanticist has got the nonexistents in his ontology and is stuck with them. Perhaps someone may find a way of doing modal semantics *without* genuinely (objectually) quantifying over mere possibilia; such a semantics would be welcome. But until it is produced, Quine contends, the modal theorist is committed to the presence of "nonexistent" possibles in his official ontology.[12]

So much the worse, Hintikka (1969) remarks, for Quine's vaunted criterion of ontological commitment.

We have to distinguish between what we are committed to in the sense that we believe it to exist in the actual world or in some other possible world, and what we are committed to as a part of our ways of dealing with the world conceptually, committed to as a part of our conceptual system. The former constitute our ontology, the latter our 'ideology.' What I am suggesting is that the possible worlds we have to quantify over are a part of our ideology but not of our ontology.... Quantification over the members of one particular world is a measure of ontology, quantification that crosses possible worlds is often a measure of ideology. [p. 95]

Thus, an item's simply being the value of a bound variable is not sufficient for that item's being genuinely posited or for its being a member of our ontology. So, Hintikka concludes, Quine's criterion of ontological commitment should be rejected as incorrect and replaced by a suitably restricted criterion.

I believe this contention betrays a misunderstanding of Quine's use of the backward-E test as revealing a theorist's commitments. Hintikka maintains that Quine's view, or his usage, is *wrong* and needs to be revised. But, as I understand him,[13] Quine never intended his "criterion" to be in any way substantive or controversial. Insofar as he regards it as correct, he takes it to be trivial: Writing the backward E is the logician's way of making an existence claim and officially adding something to his ontology. That is just what a quantifier *is*, at least on its ordinary interpretation. Thus, if Hintikka wishes to continue in this vein, he will have to show Quine some other, nonstandard way of understanding the quantifiers, and that would be precisely to concede Quine's point. Let us therefore return to our consideration of possibilia and parsimony.

It might be complained that Quine's objections to possibilia are purely *aesthetic* in character; Occam's Razor itself is an aesthetic principle. But, according to Quine, this feature of his reason for rejecting Meinong's ontology is shared by virtually every reason anyone ever has for rejecting or accepting any theory of anything (particularly when the subject matter is highly abstract), and any epistemological Explanationist should agree. Meinongian quantification must therefore be justified on the basis of its explanatory value.

Cresswell (1972), Stalnaker (1976), Lewis (1986), and others have appreciated this and argued that possible worlds and their denizens are rich in explanatory utility. In fact, it seems, we have no idea of how to go about doing modal semantics, decision theory, Montague Grammar, or any such thing without quantifying over possibilia; it is hard even to imagine what a competing approach would be like. For now, therefore (it is claimed), we *must* quantify over possibilia no matter what further aesthetic objections Quine has brought against them. Nor do we violate Occam's principle in doing so, for that principle only forbids our positing entities *beyond* explanatory necessity.

Quine surely sees the force of this argument; but refuses to grant the premise, for any number of fundamental reasons: (a) He believes that most of the "data" for modal and for proposition-attitude semantics are unreal or negligible (e.g., his philosophical skepticism about "necessity" is well known). (b) He has argued tirelessly that syntax and "semantics" are indeterminate; he holds that linguistic semantics of the sort that is currently in favor is a pseudoscience and a pipe dream in any case.[14] (c) He holds that "explanations" of the sort provided by possible-world explications of modalities are not explanations in any genuine sense. To say that "It is possible for there to be pink elephants" is true in virtue of there being some possible world in which there are pink elephants is to offer a pseudo- or "dormitive virtue" explanation.[15]

Here we soon arrive at a set of basic methodological differences between Quine and the possibilia enthusiasts, having to do with the nature of science, the nature of explanation, and so on. Perhaps then it is a fundamental impasse in this area that explains the intractability of the dispute between the friends and the foes of possibilia. And yet I do not think it is. I shall now argue that the real problem lies elsewhere and has little or nothing to do with Occam's Razor; in the rest of this Part, I shall try to make some progress toward its resolution.

3. THE REAL PROBLEM

It is maintained that, no matter how alien, ugly, and awkward nonexistent possibles may seem, they satisfy an aching theoretical need and that we have no real choice but to posit them as we have posited any other odd kind of abstract entities for explanatory purposes. I believe that this represents a bad misconception of the issue.

When we posited properties, some philosophers complained that properties were queer and obscure. When we posited sets, other philosophers complained that sets were queer and obscure; and the same for propositions, negative facts, and so on. But in each of these cases it was fairly clear what was being said, even if it was hard to imagine how an object could be nonspatiotemporal and yet be apprehended by a human mind or whatever; we understood the *quantificational* part of quantifying over abstract entities, even if we did not understand the nature of the entities themselves.[16] And this is what distinguishes nonexistent possibles from all the foregoing kinds of posited abstract entities: A first try at quantifying over "things that do not exist" yields

(A) $(\exists x) \sim (\exists y)(y = x)$

(cf. Meinong's "paradoxical" formulation [1904/1960, p. 83]). And this formula is a *contradiction*. The crux is that, unlike the notion of a property or a proposition, the notion of a *nonexistent* thing or world is not merely queer or obscure or marginally intelligible, but is an apparent overt self-inconsistency.

Of course, no friend of possibilia intends the view to be understood in this absurd way. It is at this point that many different moves can be made.

I suggest that the best way of organizing the whole issue is to see the various contemporary metaphysical theories of possibility as being partially or wholly conflicting ways of resolving the prima facie contradictoriness of (A). For this is the first job to be done; I cannot see that any further talk of theoretical utility is germane for now.

4. CONTINUING HOSTILITIES

The obvious first move in this new direction is to disambiguate (A)'s quantifier. Meinong might therefore distinguish two different operators, one continuing to indicate actual existence and the other indicating some so far mysterious *secundum quid*. The former would be given the usual model-theoretic semantics; the latter would remain to be explained.

Meinong and other apologists for possibilia point out that this "mysterious" second operator already has a perfectly intelligible and straightforward English counterpart, viz., that which occurs in "There are things that don't exist," "There is a character in *Hamlet* who is smarter than anyone in our department," etc. Undeniably such constructions occur in true sentences of English, and *almost* undeniably they occur in true sentences of English at that. The Meinongian may now say that insofar as we understand the casual use of such expressions, we can understand (A)'s leftmost quantifier--as a translation of this sort of thing. And this quantifier, introduced into our logical theory and into our semantical metalanguage for English, will bear the whole weight of our possibilistic apparatus. For our original, actuality-indicating quantifier can easily be defined in terms of our Meinongian quantifier, but not vice versa. "Actual" and "existent" figure as predicates in Meinongian English, and it is simple just to take this usage over into our logical theory. Thus, "There are things that don't exist" would be translated into Meinongian, not as (A), but as

$$(\exists x)_M \sim \text{Actual}(x)$$

or as

$$(\exists x)_M \sim (\exists y : \text{Actual}(y))(y = x),$$

neither of which is formally contradictory. And our original standard quantifier can now be introduced as a defined sign:

$$(\exists x)_A \ldots x \ldots =_{df} (\exists x)_M (\ldots x \ldots \& \text{Actual}(x));$$

or

$$=_{df} (\exists x : \text{Actual}(x))_M \ldots x \ldots$$

Disambiguated accordingly, (A) would be ruled satisfiable by whatever semantics the Meinongian intends to provide.

This last nonspecific referring phrase was tokened in a smug and deprecating tone. It presupposes a demand on our part that the

Meinongian search for, find, and offer "a semantics for" the quantifier "$(\exists x)_M$," and we have some fairly specific constraints in mind as to what is to be counted as success in this. In particular, what I am implicitly demanding is a *model-theoretic* semantics, done entirely in terms of actual objects and their properties--for what else *is there really?* I am allowing the Meinongian the funny operator "$(\exists x)_M$" only on the condition that it be explained to me in non-Meinongian terms.[17]

To this the Meinongian may reply that it is child's play to give a model-theoretic semantics--one whose domains include nonactual objects, true enough, but that is all right, since *there are* nonactual objects, after all. And so it seems we have arrived at another impasse.

There is no formal circularity in this last Meinongian move. The Meinongian is explaining in the meta-metalanguage (English) how the expressions of the Meinongian semantical metalanguage are to be understood, and so is not simply defining a linguistic item in terms of itself. But, clearly, anyone who *really needs* a semantics for "$(\exists x)_M$" (that is, anyone who needs a semantical explanation in order to understand "$(\exists x)_M$" *at all*) is not going to be delivered from distress by an appeal to--or rather, by unexplained use of--the very term of the meta-metalanguage of which it was introduced as a translation. The Meinongian evidently wants to take the nonactualist "There is" *and* its canonical counterpart as primitives of their respective containing languages.

The difficulty that lies between the Meinongians and the philosophers who sport a "robust sense of reality" can thus be understood as being an intractable difference in what they are willing to take as primitives. Quine takes just his standard quantifier as primitive (and likewise talk of "existence," "actuality," etc. in English), and would seek some actualistic regimentation of Meinongian constructions in English. The point of such a regimentation would be to enable us to understand the Meinongian constructions, in terms of other constructions that we understand antecedently.[18] By contrast, the Meinongian takes "There is" and $(\exists x)_M$" as primitive and explains the actualistic usage in terms of them. These two competing choices of primitives may engender a pair of mutually incomprehensible conceptual schemes.

14   CHAPTER ONE

It should be noticed that the impasse is slightly lopsided: Though each of the participants has so far run up just one unexplained primitive, the Meinongian is in fact stuck with a second, viz., the predicate "Actual." There is no obvious way of explicating "Actual" in terms of "$(\exists x)_M$" plus notions accepted by both sides (that is, not without the use of further new primitives), though in the next chapter we shall discuss an unobvious way. It is true that, so far as has been shown, attempts to explicate the Meinongian "There is" in terms of the actualistic notions already in play in reality-oriented semantical metalanguages may just fail and that we *may* therefore be forced to introduce a new primitive into the metalanguage amounting to a Meinongian quantifier. But that is what remains to be seen. The literature now contains a surprisingly various array of theories of possibility, which I shall now begin to distinguish.

5. APPROACHES TO POSSIBILITY

Again, I take the first task of any philosophical or semantical theory of possibility to be to solve the prima facie contradictoriness of Meinongian formulations such as that translated by (A). I shall distinguish five basic approaches. They are not exclusive; some current views fall into more than one of the five categories. They may not turn out to be exhaustive either, though it is hard to imagine a further alternative.[19]

(i) The *Relentlessly Meinongian* approach. This is simply to leave our impasse as it is, embrace the Meinongian's two primitives, and dismiss Quinean-Russellian hostility as perverse.[20] A Relentlessly Meinongian theory must be elaborated and ramified in something like the ways I have suggested in section 1; such elaborations in fact have been carried out in some detail.[21]

(ii) The *Paraphrastic* approach. Some philosophers have believed that apparent reference to and "quantification over" nonexistent possibles could be eliminated by contextual definition, i.e., paraphrased away from whole sentences in which they occur. Possible individuals and possible worlds would then be treated as *façons de parler*.

Perhaps under the Paraphrastic heading should be included *fictionalist* approaches to possibilia. Only one such is in the field, that of Armstrong (1989), which will be discussed in Chapter 3.[22]

(iii) The *Quantifier-Reinterpreting* approach. A practitioner of this method attempts to meet our Quinean challenge directly and provide a nonstandard semantics for the Meinongian quantifier which preserves its inferential properties but requires no nonactual entities.

(iv) The *Ersatz* approach, as it is called by Lewis (1986). An Ersatzer leaves both quantifiers standardly interpreted, but construes "possible worlds" etc. as being actual objects of some kind. The Ersatzer's ploy is to find some actual entities which are structurally analogous or isomorphic to an adequate system of possible objects and worlds, and which therefore can *do duty for* or *serve as* possibilia; he/she may then let the apparently Meinongian quantifiers range over these objects and define the Meinongian's "Actual" in terms of some property that some of them but not all of them have. Ersatzer positions differ, naturally, according to what objects they take as ersatz worlds or world-surrogates; "other possible worlds" have been construed as sets of sentences or propositions, set-theoretic reorderings of the basic elements of our own world, types of mental act, and more.

(v) David Lewis' (1986) view, which like the Meinongian countenances nonactual entities but which defines, not the Meinongian quantifier, but the narrow actuality quantifier.[23] Of this, more shortly.

I shall understand the term "Actualism" to cover approaches (ii)-(iv) and any other position that deflates or apologizes for Meinongian quantification and refuses to monger nonactual entities. For want of a more perspicuous name, I shall follow van Inwagen (1986) in calling (i) and (v) "Concretist."[24] The "Actualist"/"Concretist" division has become tolerably clear within modal metaphysics, especially as applied to worlds in particular.

Actualism is a "one-world" construal of logical space. The actual, blooming, buzzing physical world of earth and fire and iron and concrete and flesh and blood is world enough; the "other, nonactual possible worlds" invoked by intensional logicians are only abstracta, either abstract objects or mental constructs or some sort.

Concretism is based roughly on Meinong's doctrine of *Sosein*, according to which every possible world--not just the favorite "real" one that we inhabit--is physical, made of earth or fire or ... or combinations of those as the case may be. On Lewis' picture in particular, "other possible worlds" are in no way merely abstracted or constructed, or otherwise ontologically subsidiary to our own world.

I shall postpone discussion of Ersatz theories until Chapter 3, and the critique of Concretism until Chapter 4. In what remains of this chapter I shall address the Paraphrastic and Quantifier-Reinterpreting approaches, and sketch my reasons for thinking that neither method offers a promising start toward the explication of possibilistic talk. Then in Chapter 2 I shall survey the Actualist/Concretist conflict in finer detail.

6. THE MEINONGIAN QUANTIFIER AS NONOBJECTUAL

The Paraphrastic and the Quantifier-Reinterpreting approaches share a tenet: that our Meinongian quantifier "$(\exists x)_M$" is not what at first it appears to be. On the former approach, the "quantifier" functions only as part of an idiom or *façon de parler* and is not really a quantifier at all. On the latter, "$(\exists x)_M$" remains a quantifier at least in the minimal sense of preserving "$(\exists x)_M$"'s standard implicational relations and receiving its own base clause in a Tarskian truth theory for the Meinongian language, but the truth theory in question characterizes it in other than its usual objectual way.

An obvious instance of the Paraphrastic strategy is Russell's own treatment of possible objects in "On Denoting" (1905b) which falls cleanly out of his Theory of Descriptions. This version has well-known disadvantages. One is that virtually all the sentences "about" possibilia that Meinong would regard as straightforwardly true come out false on Russell's proposal, precisely because of the nonexistence of the possibilia in question. Another is that Russell's method of handling problems of intensionality generally has not proved to be powerful enough to yield a satisfactory systematic treatment of those problems.[25] Finally, the method offers no obvious way of eliminating talk of possible *worlds* from modal semantics, which would be our main task.

A more promising Paraphrastic program would be to understand "possible-world" talk counterfactually, as has been suggested by Kripke.[26] This is quite a natural suggestion and does much to make talk of possible worlds more homey. An antic sentence such as "In some possible world distinct from our own, Richard Nixon is a Black Panther" might be paraphrased as "Had things been otherwise, Richard Nixon might have been a Black Panther," a sentence which we all more or less understand or at least would not balk at in ordinary conversation.

The counterfactual approach is inadequate in two serious ways, I think. First: It is not enough to provide a sample paraphrase or two. The counterfactual theorist would have to work out a systematic and rigorous *formula* for paraphrasing formal, model-theoretic sentences concerning possible worlds, and in such a way as to preserve all the theorems of our logical theory and all the advantages of each of the modal logics or modal semantics under analysis. It is hard to imagine how this would go.[27] The difficulty becomes critical when we not that any adequate modal semantics will require many *sets of* possible worlds, sets of sets of worlds, and so on.[28] The counterfactual approach is not allowed to leave set abstraction on worlds undefined. (This seems to me to be a crucial point, one that I have never heard a Paraphrastic theorist address.) Even if we have provided a satisfactory system of eliminative contextual definitions for *quantification* over nonexistent possibles, this system would have to be extrapolated to cover set abstraction as well, and no way of doing this in terms of counterfactuals comes to mind. (We might try invoking "ways things might have been" and abstracting on them, but to do that would be to reify the "*ways*" and leave us with all the same problems we had before.)

My second complaint about the counterfactual approach (reprised from Lycan and Nusenoff [1974]) is this: Granted, we "understand" counterfactuals in ordinary conversation. But for purposes of serious philosophy they have proved to among the most troublesome and elusive expressions there are. Their truth-conditions have remained genuinely (not just officially) mysterious; their well-known context-dependence has not been understood at all; and in a discussion or seminar on conditionals, people blank out or disagree even on very basic matters of data. Resting a philosophical theory on unexplicated counterfactuals is like hoping one may cross a freezing river by hopping across the heaving ice floes. Great progress has been made on the general understanding of counterfactuals in the last twenty-five years, largely through the work of Stalnaker and Lewis; and the source of this progress is the considered, ingenious, and well-motivated use of possible-worlds semantics. Therefore we have extremely strong reason to analyze and understand counterfactuals in terms of possible worlds. But if we are to do so, we cannot without circularity turn back and paraphrase away talk of possible worlds in terms of unexplicated counterfactuals.

I cannot say what other Paraphrastic programs might be devised concerning mere possibilia. But I think the kinds of considerations I have brought out so far give us substantial reason to look along still other lines. The same, I believe, can be said of the Quantifier-Reinterpreting approach, which has only one existing instance that I know of: Ruth Marcus has proposed[29] that Meinongian quantification be understood as *substitutional* quantification.

This suggestion has a good deal of intuitive appeal. It is quite plausible in the case of "There are things that don't exist": when I utter that sentence aloud, I feel a tendency to continue by listing true negative existentials ("...you know: the round square doesn't exist, Macbeth doesn't exist, the free lunch doesn't exist, and so on"). And there certainly is plenty of overt quantification in English that is substitutional. It might be thought that Marcus' proposal fails in the case of possible worlds on the grounds that worlds in general do not have names at all (the substitution class would be far too small); but we may easily generate a system of canonical names for possible worlds from existing resources: Each world, we may suppose, is correctly described by a maximally consistent set of sentences $\{P, Q, R, ...\}$; to obtain a name of a world in which P, Q, R, ..., simply form a definite description from the latter indefinite one: "$(\iota w)\text{In}_w(P \& Q \& R \& ...)$."

This proposal will inherit the usual sorts of problems that philosophers have raised for substitution interpretations of more familiar quantifiers. The most obvious of these is the "not enough names" problem: First, given that almost any real-valued physical magnitude characterizing our world will have nondenumerably many nonactual worlds corresponding to it, the cardinality of the set of all worlds--if that notion is not undefined or paradoxical--will be inconceivably high. But there are only denumerably many names of the sort exemplified above; thus, it seems, universal quantifications over worlds will be verified more easily than we would like.[30] Second, a number of philosophers have argued that the substitution interpretation somehow collapses into the standard interpretation when incorporated into a full-scale truth theory. Kripke (1976) refuted at least the most salient versions of this charge, though I shall argue in Chapter 9 below that there is a further, somewhat related difficulty which impugns the usefulness of the substitution interpretation for the truth-conditional analysis of *natural* languages.

As I see it, the main problem for Marcus' proposal is the same that arose for the counterfactual approach: How will Marcus reinterpret set abstraction and all the other operations that will need to be applied to names of "worlds"? (Certainly there have been attempts at metalinguistic reinterpretations of set abstraction, but they have concentrated on reinterpreting the abstractor itself, not the variable it binds; that is, they have concentrated on detoxifying mention of the *sets* in question, not mention of the sets' members.)

If I am right in thinking that we should continue to look elsewhere for an Actualist understanding of possibilistic talk, only one option remains: to seek an Ersatzer analysis. That will be the work of Chapter 3.

NOTES

[1] The idea that when worlds are introduced as the values of variables, the apparently distinctive inferential properties of the alethic modalities fall right out of the familiar inferential properties of the quantifiers, and the concomitant identification of necessity with truth in all possible worlds, are now so commonplace that we easily forget how stunning the idea is. It is almost universally credited to Leibniz, but I know of nowhere that it appears in Leibniz's standard texts--despite casual allusions to the doctrine, and even some (specious) page references, by a number of commentators. The most suggestive passage I have been able to find is this one: "Hinc jam discimus alias esse propositiones quae pertinet ad Essentias, alias vero quae ad Existentias rerum; Essentiales nimirum sunt quae ex resolutione Terminorum possunt demonstrari; quae scilicet sunt necessariae, sive virtualiter contradictorium. Et hae sunt aeternae veritatis, nec tantum obtinebunt, dum stabit Mundus, sed etiam obtinuissent, si DEUS alia ratione Mundum creasset" (Couturat, *Opuscules*, PHIL., IV, 3, a, 1, p. 18). I am indebted to Michael Hooker and Robert Sleigh for the reference.

[2] See Lewis (1972), Stalnaker (1972) and Cresswell (1973) for clear and well-motivated explications of this connection.

[3] Cf. Hintikka (1972, fn 6).

[4] See Cresswell (1990).

[5] He does not declare himself on the exact felt ground of his firm belief in possibilia. I suppose the intuitive irresistibility that his Objects had for him is best explained by ascribing the appropriate semantical views to him. The irresistibility in turn of a naïve semantics for superficial singular terms is due to there having previously been no competing *semantical* theories such as Russell's to weigh against the naïve account. (This is not to say that he would have gone on to accept Russell's account if he had thought of it.)

[6] Russell (1905/1956), Quine (1948/1963).

[7] Similarly, Lewis (1973, sec. 4.1) began his discussion of concrete possible worlds with a brief paean to the hominess and familiarity of nonactual worlds. I shall argue below that his "natural as breathing"

talk, like Meinong's, thinly masked a formidable theoretical apparatus which must be evaluated on theoretical grounds. Lewis (1986) tore off the mask, in any case.

[8] Meinong relies on this principle in displaying the independence of *Sosein* from *Sein*: the round square, he says, must at least be round and be square.

[9] In fact, Routley and Routley (1973) remind us that Meinong did go on to impose a type theory of this sort.

[10] Forrest (1982) has pressed a related question involving skepticism; Lewis (1986, sec. 2.5) replies at some length.

[11] What is to prevent our positing any crackpot kind of abstract entity we like and immunizing ourselves against criticism by adding that the entity is nonactual? For that matter, what is to prevent our positing lavish supplies of *physical* things such as unexplored planets and doing the same? Routley (1980, pp. 755ff.) does precisely the former, and even extends the treatment to past and future physical things.

[12] Notice, as a referee has reminded me, that the issue of ontological commitment is independent of the charge of bloating or jungle. Meinong might resolve his terminological differences with Quine and still disagree as to what one's ontology should include.

[13] Cf. Harman (1968, p. 348).

[14] See, e.g., Quine (1972). I discuss the impact of Quine's views on linguistics in Chapter 9 of Lycan (1984). N.b., if for some reason you do not already own a copy of that book, it would be a good idea to go and buy one.

[15] Cf. Harman (1967) on the positing of intensional entities in semantics. On "dormitive virtue" explanations, see Chapter 10 below.

[16] Of course, there are philosophers whose revulsion for intensional entities Platonistically understood is so great that they have tried to *reconstrue* such quantification as not having its normal meaning. Rudolf Carnap was one; Wilfrid Sellars another.

[17] Wallace (1972) seems to propose that we *can do no other* than to make such a demand vis-á-vis any canonical idiom if we are to understand that idiom.

[18] I am here glossing over Quine's insistence that "regimentation" (the replacement of an awkward or troublesome locution in a natural language by a formal construction of some chosen canonical idiom for technical purposes) does not do anything that might properly be regarded as displaying or even illuminating the *meaning* of the original locution. This denial falls trivially out of Quine's indeterminacy doctrine. So he would balk at my talk of offering an Ersatzist "semantics for" the Meinongian quantifier in the going sense of the term (cf. fn 13 above). For purposes of this book I shall waive that issue and continue to take semantical theories to be genuine theories about their target locutions. I have discussed Quine's indeterminacy doctrine in Chapter 9 of Lycan (1984).

[19] I shall not here consider nonreductive approaches to modality, i.e., approaches that leave model operators primitive rather than trying to explain their behavior by explicating them in terms of worlds (e.g., Prior and Fine [1977], Forbes [1985], Zalta [1983, 1988]).

[20] Hereafter, I shall use the bare adjective "Meinongian" to designate just this two-primitive position. Being "Meinongian" in this sense does not require acceptance either of "incomplete" objects or of impossibilia.

[21] Castañeda (1974, 1989); Parsons (1974, 1980); Rapaport (1978); Routley (1980). Zalta's (1983, 1988) ontological apparatus of "Abstract Objects," when given a Meinongian interpretation, is the best worked-out such view I know, but Zalta himself does not favor the Meinongian interpretation.

[22] To my knowledge, no one has ever suggested a plain Error theory in the sense of Mackie (1977). It is easy enough to see why; if there simply are no modal or other intensional truths, then English contains virtually no true sentences.

[23] That work was preceded by Lewis (1970) and (1973, sec. 4.1). See also Unger (1984).

[24] That term has two serious drawbacks. First, it suggests that every possible world is concrete and altogether lacks abstract elements of its own; for all we know, there may be (or "be") possible worlds consisting of nothing but abstract entities such as numbers. Second,

the "abstract"/"concrete" distinction itself is seriously vexed; see Lewis (1986, sec. 1.7), and Hale (1988).

[25] Smullyan's (1948) well-known attempt to extend Russell's theory in this way is criticized by Linsky (1968). I tried to push Russell's treatment in a slightly different way in Boër and Lycan (1975, sec. II).

[26] Kripke (1971, 1972), and particularly in conversation. Rescher (1975) shows considerable sympathy for the counterfactual approach, though he does not adopt it in the end. I do not mean to imply that either Kripke or Rescher has made any attempt to work out the approach; so the objections I shall be raising against it are not directed at the writings of any actual person.

[27] The point I am making here is very easily overlooked by philosophers seeking paraphrastic eliminations of the metaphysically dubious entities assumed by some formal and highly technical theory. Consider a simple (and mythical) example: It seems that sentences of the form "$x \; \varepsilon \; y(Fy)$" can be paraphrased simply as "Fx," as Quine has observed. Thus, some class abstracts may be regarded as *façons de parler*. A philosopher who lacked Quine's own mathematical sophistication might well come to think that nominalism had been achieved, in that Quine had hit upon a program where class abstraction and talk of classes could be paraphrased away. What this naive philosopher would be overlooking is that simple class abstraction is not the only technical operation that occurs essentially in set theory. In this case, as Quine points out, one would not even be able to explicate talk of classes of classes, since his "virtual class" device leaves undefined any construction in which a bound variable occurs immediately to the right of "ε." A nonmythical example of this optimistic sort of fallacy is some philosophers' reaction to Wilfrid Sellars' approach to abstract entities. Sellars, mobilizing his ingenious device of dot quotation, has offered some very plausible paraphrases for simple talk of properties, propositions, sets, and so on. Philosophers justly impressed by the cleverness and by the naturalness of these paraphrases have taken Sellars to have offered an acceptable nominalistic *theory of* abstract entities generally. But Sellars has given us no reason at all for thinking that the rarefied operations of (e.g.) graph theory, the integral calculus, differential geometry, or other

areas of higher mathematics can be explicated in terms of dot quotation, since he has provided no directions in which his original paraphrases of simple and relatively nontechnical constructions are to be extrapolated. (Sicha [1974]) has tried to provide at least one such direction, but the mathematics he is able to treat is very elementary indeed.)

[28] Perhaps the most elaborate set-theoretic world-encapsulating edifice to be found in the existing literature is that built up by Cresswell (1973).

[29] Another option may emerge from Rantala's "urn" semantics; see Rantala (1975) and Hintikka (1975).

[30] Several philosophers have recently claimed that the "not enough names" problem is soluble for the usual sorts of subject matters such as arithmetic. See (e.g.) Sicha (1974). These purported solutions have not been thoroughly adjudicated in the literature, so I cannot say whether they could correctly be extended to cover the cardinalities involved in a system of nonexistent possibles. Lewis (1986, sec. 3.2) suggests that linguistic Ersatzers make use of "Lagadonian" languages in which things serve as their own names; but a proponent of the Quantifier-Reinterpreting approach to our problem can hardly propose to let actual possible worlds serve as their own names.

CHAPTER 2

THREE CONCEPTIONS OF POSSIBLE WORLDS

Once it has been settled (so far as it has been settled) that quantification over mere possibilia is literal and objectual, one must choose between an Actualist reinterpretation of the possibilist quantifier and a more outrageous Concretist interpretation. Surprisingly, even within Concretism one must choose between a genuinely Meinongian view (such as that of Richard Sylvan *né* Routley) and David Lewis' unexpectedly different position.
My purpose in this chapter is to explore the Ersatzer/Concretist differences further, bringing out some points that have gone unnoticed or unappreciated. In particular, many of the hot debates current in modal metaphysics are *almost* automatically settled once one makes one's fundamental choice between Actualism, Meinongian Concretism and Lewis' view; thus they are not independently real issues, though one's independent *preferences* in regard to them should help guide the fundamental choice.

1. A THUMBNAIL HISTORY

Interestingly, possible-worlds semantics came into *this* world by way of explicitly Ersatzist interpretations. Wittgenstein's *Tractatus* suggested what Lycan (1979) called a Combinatorialist account of worlds, the idea that "other worlds" are abstract rearrangements of what are in fact the ultimate constituents of this world.[1] Carnap with his state-descriptions was the original "Linguistic" Ersatzer in Lewis' phrase, construing worlds as sets of sentences; and in the late 1950s,

Jaakko Hintikka followed Carnap closely, speaking of "model sets" of formulas.[2]

During the ensuing period in which possible-worlds semantics matured and flourished, Meinong continued to be shunned as a madman and/or a moron misled by surface grammar; his idea that the "actual" or existent objects were only some objects among others had successfully been ostracized by Russell. Not until 1970 did Lewis challenge that Russellian orthodoxy; and not until the mid- to late 1970s did serious systematic Meinongian works begin to appear, following Meinong's (and his student Mally's) own turn-of-the-century writings.[3]

Lewis (1970), followed by Routley and Routley (1973), Castañeda (1974), and Parsons (1974), ventured back towards Meinong's view by refusing either to deflate or to apologize for the mere possibilia they invoked. Actualists such as Plantinga (1974), Adams (1974) and Stalnaker (1976a) reacted, with controlled horror, by reminding us of the virtues of Ersatzism. And so the battle was joined.

What is less clear is exactly how the battle line is drawn. The best I have said till now is (again) that Actualists believe there is only one concrete physical world, though of course there are other ways the world might have been, which "ways" can be represented by effete and filmy abstracta of some sort; while Lewis and the neo-Meinongians believe that alongside our concrete world, there are other *equally concrete worlds*, whose "nonactuality" is less obviously a matter of ontological kind. The Concretist holds that there are many genuine worlds--not just world-simulacra--of which ours is only one. Lewis adds the further claim that words like "actual" are locative indexicals on the model of "here" or "this planet": What we call the "actual" world is *this* world, the one we inhabit; denizens of other worlds are equally correct in using "actual" to refer those worlds of their own.

Even now, the difference may be thought only metaphorical, or just a difference in how we imaginatively depict to ourselves the quantificational apparatus underlying intensional logic. (Someone might say, "Well, of course the Golden Mountain is a *mountain* and not another thing, so in that sense nonactual worlds are physical and concrete, but in another obvious sense the "alternative worlds" are just intellectual creations of ours; where is the disagreement?") The main purpose of this chapter is to dispel that irenic impression by showing that the choice of Ersatzism vs. Concretism has real and substantive doctrinal consequences.[4]

Incidentally, I do not think we any longer have the Quinean option of simply ignoring the issue, either. I believe possible worlds or proxies for them are now indispensable for detailed work in semantics, just as numbers or proxies for them are indispensable to work in natural science. That is not to say that either the worlds or the numbers must exist in any ontologically ultimate sense, for either might be a *façon de parler* or otherwise a fiction. It is to say that some account must be given of them.

The main claim of the previous chapter was that the root difference between Actualism and Concretism has to do with quantification. In particular, the Ersatzer leaves quantification itself as is, while the Concretist is forced to distinguish two contrasting uses of the quantifier. I count that as a severe initial handicap for the Concretist, but Lewis (1986) has taken a novel and significant step towards mitigating it.

2. LEWIS' REVERSAL

How then are we to understand the more inclusive Meinongian quantifier "$(\exists x)_M$"? We may simply assume that we understand the narrow, realistic quantifier and take words like "actual" and "real" as primitives, but the Meinongian asks us to take the more inclusive quantifier as primitive also. (An Ersatzer, of course, has no trouble defining it.) Some Meinongians (notably Parsons [1980] and Routley [1980]) are entirely unembarrassed by the extra primitive; but it seems to me both a grave theoretical liability and an obstacle in practice to making Meinongianism palatable to very many philosophers.

Prior to reading Lewis (1986), I had been able to imagine only three possible ways of avoiding the Meinongian liability, i.e. (so I thought), of defining the inclusive quantifier rather than leaving it primitive: as before, Paraphrase, Quantifier-Reinterpretation, or

Ersatzing.) But in *On the Plurality of Worlds*, Lewis reverses the Meinongian problematic, and thereby adds our fourth possible way of distinguishing the two quantifiers without simply announcing that they differ in meaning and then telling the audience to shut up. As we shall see, this innovation has both a stunning advantage over its Meinongian predecessor and a sobering drawback.

Lewis' trick is to define, not the inclusive quantifier, but *the narrow Actualist quantifier*. (That is the taxonomic possibility I had failed to anticipate.)

On Lewis' Mature view, as I shall call it, the more inclusive "Meinongian" quantifier is not Meinongian at all, but just the ordinary quantifier, and has its ordinary, everyday, (usually) physical meaning; it needs no explication or interpretation whatever. Instead, Lewis defines the narrow quantifier. It is of course understood as the inclusive quantifier restricted to the actual, but now Lewis introduces the novelty: He goes on to explicate "actual" in terms of the concept of a *worldmate* (actual things are worldmates of *ours*), and he offers an independent account of worldmateship in turn, as spatiotemporal relatedness.

The latter account is independently motivated. For the Concretist has a problem that has never troubled the Actualist: What distinguishes one world from another? According to the Actualist, worlds are simply different *sets* or other actual items. But since Lewis holds that he and you and I are *physical* worldmates while neither Polonius nor the Wife of Bath is a worldmate of ours and neither are they worldmates of each other, Lewis must specify what it is in virtue of which two things are worldmates or not. This is a nontrivial problem,[5] and Lewis' own solution is to suggest that what separates one world from another is simply *spatiotemporal disconnection*: items in one world bear no determinate spatiotemporal relations to items in any distinct world.

...things are worldmates iff they are spatiotemporally related. A world is unified, then, by the spatiotemporal interrelation of its parts." (1986, p. 71)

I shall discuss that suggestion further below, but the point for now is that, assuming worlds are individuated in that way, worldmateship can then easily be defined as spatiotemporal relatedness.

The Lewisian upshot is that: Every possible world exists, in the perfectly ordinary sense of "exist"; only one world is actual; actual things are just those which are spatiotemporally related to you and me; and at any world, the word "actual" refers to the things of that maximally connected region, not to the things *we* here at @[6] call "actual."

The strategy of defining the narrow quantifier instead of the inclusive one completely removes the (to some) incomprehensibility of standard Meinongianism. What Lewis is now saying is absolutely, transparently clear. For Mature Lewis, a statement of possibility tacitly quantifies over ordinary physical objects that have what may be bizarre physical properties. E.g.: The innocent modal sentence "There might have been talking donkeys" is held to be true iff there exists (in the absolutely everyday sense of "exist") at least one donkey (in the absolutely everyday sense of "donkey") such that (a) it talks, and (b) it is spatiotemporally dislocated from us.

This complete removal of unclarity is the "stunning advantage" I mentioned earlier. The Mature theory's "sobering drawback" (ditto) is its *outrageous falsity* (of which more in Chapter 4). But as Lewis often reminds us, everything comes at a price.

Notice that relative to the Meinongian, ironically, Lewis is himself in the position of an Ersatzer. Where the Meinongian has posited weird "nonexistent Objects" in the range of an unexplicated *and* who-can-say-what-cardinality-inclusive quantifier, Lewis hygienically replaces the weird Objects by good old-fashioned (if in some cases ill-behaved) physical objects and the offending unexplicated quantifier by our familiar existence-claiming one. (He differs from the Actualist Ersatzer only in denying that his Ersatz Objects are actual and in making them physical rather than abstract or mental.) We can imagine the Meinongian reproving him for his grab at "paradise on the cheap," for trying to get by with just the one quantifier eked out with what are merely existent simulacra of genuinely nonexistent Objects.

Incidentally, it is high time for a word about an old exegetical dispute. In Lycan (1979), on which my present Chapters 1, 3 and 4 are based, I interpreted Lewis as a Meinongian. This was for two reasons: Lewis had never in print distinguished himself from the Meinongians (bar his rejection of "incomplete" individuals and impossibilia), and in particular he had said nothing of any independent explication of worldmatehood in terms of spatiotemporal connectedness. I was soon reviled by Routley (1980, pp. 494-5, fn 1) for construing Lewis as a Meinongian--an odd precognition, for whatever Routley knew in 1980 that I did not, he did not find it in Lewis' written works. Indeed, no one could have found it anywhere, until someone had formulated a proposal for explicating the term "actual" as applied to concrete worlds, which Lewis did for the first time in (1986). Make no mistake: Any theorist who insists that the possibilist quantifier and the existence- or actuality-restricted quantifier or predicate differ in meaning, but explicates neither of them, is a Meinongian. It is the invoking of spatiotemporal dislocation, *and only that*, that makes the difference between Mature Lewis and a Meinongian Lewis as I am using the term.[7]

3. THE INDEPENDENCE OF THE INDEXICALITY THESIS

Stalnaker (1976a) pointed out that Lewis' overall view actually contains each of several theses that are mutually independent. In particular, he argues, "...the *semantical* thesis that the indexical analysis of 'actual' is correct can be separated from the metaphysical thesis that the actuality of the actual world is nothing more than a relation between it and things existing in it" (p. 69). Stalnaker goes on to defend the indexical analysis while rejecting Lewis' metaphysical thesis. But it is worth noting that both claims are strictly independent of Lewis' Concretism as well: One *might* accept Concretism but neither the indexical analysis nor the metaphysical thesis. Let us consider in more detail the notion of a *privileged* or *designated* world.

On the Ersatzer's Actualist picture of the modal plenum, it is *axiomatic* that amongst all the "possible worlds," one predominates. What it is for a "way things might be" to be the way things actually are is for that abstractum to correspond to the one concrete, flesh-and-blood-y real world; the other "possible worlds" (that correspond only imperfectly) misrepresent or otherwise diverge from that genuine, concrete world in one respect or another. Thus the abstract "way things might be" that is actual--the way things *are*--is ontologically sharply set off from all the others.

As I have said, the Concretist could without contradiction sustain the idea of a designated world. Lewis needs to argue against that idea. But the argument is not hard to come by.

The suggestion that one amongst all the possible worlds predominates is fairly fanciful. All the worlds are genuine, concrete, blooming, buzzing physical worlds, not abstracta or intellectual constructs. It is as if we live on a planet in a solar system, amid other planets inhabited by other beings like or unlike us as the case may be. Our world is "actual" or "real" to us, but the other beings' worlds are just as "actual" or "real" to them, and seemingly in just the same sense. It is hard to resist, and Lewis notoriously does not resist, the egalitarian indexical thesis.

Again, a Concretist *could* defend a "privileged world" view rather than the indexicality thesis. But a story would have to be told to explain why and how just one of the infinitely various concrete worlds that there are acquired its uniquely designated status. Someone might hold, e.g., that from amongst all those concrete worlds, God chose just one to actualize, just one upon which to smile. This Leibnizian view sits very well with Ersatzers, who would imagine God contemplating all the various world-*blueprints* and deciding which to actualize, i.e., which to bless as determining the one concrete world that there is going to be. But the idea of God's choosing one world from among an array of other *concrete* worlds is very strange, for each of two reasons:

First, what does "actualize" then mean? For the Ersatzer, God "actualizes" a "possible world" (= a blueprint) by constructing a genuine, concrete *world* corresponding to that particular specification. But for Lewis, God sees rank upon rank of already concrete worlds.[8] What would it be for Him to proclaim one "actual"? What would actuality come to, over and above the mere fact of God's having stopped by it and uttered the *word* "actual" in a proclamatory manner?

Belying its rhetorical tone, that question does admit of a substantive and interesting answer, for actuality and *truth* go hand in hand. In fixing one particular world as actual, God would determine a detailed distribution of truth-values over all the propositions there are. The facts of the designated world would be the truths; each of the other worlds' inevitable divergences from the designated worlds would be a falsehood, "true" only "at" a nonactual world.

But that brings us directly to the second strange result of imposing the "designated world" view on Lewis' Concretist picture: Members of the undesignated worlds would believe, falsely, that they and their friends and possessions were actual, even though they simply are not. Worse,[9] they would believe that all sorts of things really happened, when those things did not really happen, but happened only in their own, objectively nonactual worlds. And if a Descartes inhabits such a world and argues there that God is no deceiver, he is cruelly mistaken, for God has wantonly robbed that Descartes of truth by actualizing a different world, and knowingly falsified many of his entirely natural and reasonable beliefs.

Notice, incidentally, a further peculiarity of this idea. Our otherworldly Descartes and his worldmates have all those justified false beliefs, but without the slightest defect in either their cognitive structures or their perceptual apparatus. Our usual Cartesian victims are all impaired in some way; even the Evil Demon or his contemporary equivalent, the mad brain scientist, has to interfere with perceptual input and/or with its subsequent processing in order to deceive. Not so for God, on the "designated world" variant of Concretism. God deceives, not by His power to confuse, delude or illude, but simply by fiat, and not even *fiat lux*.

Moreover, the real Descartes' own saving defense does not even begin to help otherworldly philosophers who have the misfortune not to exist in the real world also (even granting, *contra* Lewis, that transworld identity is possible): The *cogito* fails for them! Descartes argued that not even the Evil Demon could deceive him about his own existence, since to be deceived he would have to exist and so would not after all be deceived about that. But Descartes was clearly thinking of the Demon as being a worldmate of his victim and as causally affecting the victim's mentation. True, the otherworldly philosopher cannot be deceived by a *malin genie* in his own world into thinking he exists "at" that world when he does not. But in virtue of God's having chosen to actualize a different world, *the philosopher is wrong in thinking that he actually exists*.

And this brings us back to the question of why, for that matter, inhabitants of the designated world have reason to think they are actual. They have plenty of good evidence, of course--overwhelmingly good evidence that things really are as they seem to be. But many of the poor deluded nonactual people have just as good evidence; in worlds more orderly than ours, they have *better* evidence. And they are all wrong. On the view under consideration, if you and I now are right in thinking that things actually are as they seem to us to be, we are outrageously, uncountably lucky to be right. And a strong tradition in epistemology denies that beliefs are justified when they are right only by luck.[10]

The upshot is that although it is possible for a Concretist also to hold the "designated world" view, it is hard to see why any such theorist would want to do so. Given Concretism, Lewis' indexicality thesis is overwhelmingly more reasonable.[11]

An afterword about the "designated world" view and the notion of truth: There is a familiar chicken-and-egg dispute over which concept is more basic, *truth* (period) or the relation *truth-at*. Ersatzers of course think of good plain old truth as primary, and not as world-relative--truth is correspondence to the (one) *world*--and of "truth-at" as some constructed, parasitic and abstract relation of correspondence between a sentence or proposition and another abstract object only representing a concrete world (set membership is the obvious candidate). By contrast, Concretists generally see "true" as defined in terms of "true-at": A sentence or proposition is true "period" when it is true-at the *actual* world @. Thus, for Lewis, truth comes out totally world-relative, since actuality itself is world-relative. But (here is the afterword) a Concretist who accepts the "designated world" view rather than the indexicality view of actuality will side with the Actualist one-worlder against Lewis, in the matter of truth (period) *vs.* truth-at as conceptually prior. Even in a concrete, flesh-and-blood pluriverse, truth would be realization in @, while "truth-at" other worlds would be merely "as-if" or "would-have-been" truth, entirely consistent with just plain falsity.

4. LASHED TO THE MASTS

What further issues stand or fall with the choice between our two pictures?

(I) As I have said, the Ersatzer may take the distinctness of "worlds" for granted, but the Concretist must give some account of it; and Lewis' spatiotemporal account is problematic at best. It would seem possible for a single world (a) to consist of some mutually disconnected proper parts, (b) not to have a spacetime much like ours, or (c) not to have a spacetime at all. As we shall see, Lewis speaks bravely to these objections; but considerable bravery is required.

(II) Consider counterfactuals with antecedents like "If w were actual" (e.g., "If a world in which donkeys talked were actual, donkeys would talk and so, probably, would some other species"). Such counterfactuals seem to make sense. And on the Actualist picture, they do make perfectly clear sense; but on the Concretist picture they are bizarre and need special explication (cf. van Inwagen [1980, p. 409]) In particular, if we try to explicate such counterfactuals according to Lewis' own semantics for counterfactuals, we get either nonsense (since counterfactuals about worlds tend to go undefined for Lewis)[12] or a very complicated mess: "Some world in which a world in which donkeys talk is actual (i.e., is our world) and donkeys talk and so do some other species is closer to our world than is any world in which a world in which donkeys talk is actual (i.e., is our world) and either donkeys don't talk or no other species talks."

Actually the comparison is not that simple. In one sense of "actual," that in which "actual" rigidly designates the concrete world we occupy, all such counterfactual antecedents are necessarily false, either by Ersatzer or by Concretist standards; no world distinct from this very world could be this very world, any more than Lewis could be van Inwagen. That fact shows that "actual" must also have a flaccid use. It is really there that the present comparison lies. For the Ersatzer, the flaccid use is clear: An abstract world-*story* is actual just in case it describes the one concrete world, i.e., just in case it is true, and since the concrete world could have been otherwise, a world-story that is in fact false could have been true. But it is not so straightforward for a Concretist to set out her/his flaccid use of "actual."[13]

(III) On the Ersatzer's picture, there is no special problem about individuals' persisting from world to world. If, for example, "other worlds" are big sets of some sort, actual individuals can be *Ur*-elements of those sets. But on the Concretist picture, it is hard to imagine one and the same physical individual's persisting across the boundaries that divide one flesh-and-blood physical world from another. Lewis takes that idea to be incoherent or at least mad--as if someone were to suggest that I am *strictly identical with* a rather similar but cotemporaneously existing citizen of a planet in another galaxy--and so he forsakes transworld identity in favor of Counterpart theory.[14] To do otherwise, one would have to work out a sense in which one and the same physical individual might be scattered across flesh-and-blood worlds in a way that is nonetheless consistent with the continuing impossibility of being in two distinct places at the same time.[15] It is no wonder Lewis has shunned transworld identity from the beginning.

(IV) van Inwagen (1980, pp. 419ff.) raised the problem of how we can understand the phrase "this world," since he doubted that we can uniquely *ostend* "the" world we live in. His argument assumed genuine transworld identity in addition to the Actualist picture as usual (he had previously rejected Counterpart theory); the premise is that we are citizens of more than one world at once. But even neglecting the issue of identity *vs.* counterparts, I am unpersuaded by van Inwagen's argument, for although we do inhabit many "possible worlds," there is on the Actualist picture only one *world* in the pluriverse, and we may suppose that as a default, "this world," "our world" etc. refer by deferred ostension to the only world there is.[16]

In any case, however the ostension issue is settled for the Actualist, it does not even arise for the Concretist. For Lewis, worlds are geographical regions, like planets; We can ostend our own world by pointing at anything we are able to point at.

(V) For the Concretist there is no such thing as a logically coherent but *uninstantiated* property; every property has a concrete extension somewhere in logical space. But the Ersatzer has to admit the existence of properties that are really and finally uninstantiated, even though those properties are *modelled* somewhere in the Ersatzer's mockup of the nonactual. That difference may sound small to the point of verbalness, but Lewis and Armstrong, at least, have taken it to be real and important; Lewis thinks it damaging to the Ersatzer.[17]

(VI) Formally (in model-theoretic semantics), "worlds" are analogous to *times*, in being parameters of truth; just as a property-ascription or other statement can be true at one time but not at another, so the statement can be true at one world but not at another. On the Actualist picture, this is entirely natural, since it explicates "truth-at" as set membership or something like it; different "worlds" obviously have different propositions as members. But Lewis cannot accept the analogy. That is principally because as we have seen, he must reject transworld identity, but he accepts intraworldly persistence across time (1986, pp. 202-4). If anyone can make sense of scattered but literal persistence across distinct concrete worlds, the analogy might be restored; but failing that, the disanalogy is a *prima facie* problem for the Concretist that does not even arise for the Actualist.

(VII) Once we have been disabused of Description theories of reference and various implausible qualitative essentialisms, the doctrine that Kaplan (1975) called Haecceitism is a very attractive:[18] (a) Particular persons and other things have individual essences, but those essences are nonqualitative; they are just haecceities or thisnesses, the respective properties of *being those particular things*--e.g., my own individual essence is just the property $\lambda x(x = \text{WGL})$, i.e., that of *being WGL* (himself, this very person, me). (b) Individuals are tracked through possible worlds solely as being those individuals, not by reference to any independently describable qualitative properties they have. (Each of (a) and (b) is defensible in its own right, and so each the more powerfully supports the other as well.) (c) A proposition strictly about a particular thing is a Russellian "singular proposition," one which contains the thing itself as a constituent rather than any qualitative or perspectival proxy for the thing. In terms of worlds, this is to say that what makes such a proposition true at a world is just the existence of the relevant individual in that world together with the individual's having there the right properties.

Haecceitism sits fine with the Ersatzer, who can easily allow for singular propositions and for one and the same individual *in propria persona* to be an *Ur*-element of each of many distinct "worlds." But Haecceitism is very bad for the Concretist, in part because it requires genuine transworld identity (cf. issue (III)), but also because the Concretist has no very good account to give of haecceities or thisnesses: What makes a number of other flesh-and-blood people my "counterparts" in other worlds is (to say the least) hard to specify unless in qualitative terms.[19]

(VIII) The Ersatzer is free to take Kripke's stipulative view of transworld identity:[20] (Richard Nixon might have been a Black Panther; he might also have been the third Black Panther on the right in a lineup of qualitatively similar Panthers. Indeed, he might instead have been the second Panther on the right in the same lineup in a qualitatively identical entire world. *On my sayso*, the third Black Panther on the right in world w_1 is Richard Nixon, because world w_1 is my imaginative creation and I have stipulated that we are talking about Nixon's having been a Panther instead of a conservative Republican; I am equally free to stipulate a world w_2 in which Nixon is the second Black Panther on the right, without there being any qualitative difference between the w_1 and w_2 Panthers that would make

one Nixon's counterpart at w_1 but the other Nixon's counterpart at w_2.)
But the Concretist cannot take the stipulative position, for other worlds physically exist, and their inhabitants do what they do independently of what we do here at @. Thus the Concretist must take what Kripke calls the "telescope" view, looking at other worlds as through a telescope and having to *make a qualitative case for* the identification of an other-worldly individual as a counterpart of mine.[21]

(IX) Possible-worlds semantics brilliantly translates modal and other intensional statements into the fully extensional predicate calculus. That fact enticingly suggests a total reduction of the modal to the nonmodal, the possibility of explicating the modal statements without invoking any modal primitive. But, Lewis argues (1986, pp. 150-7), every Ersatzer forfeits that possibility and is still stuck with a modal primitive: *consistency* of a set of sentences, *compatibility* of properties, or whatever. By contrast, the Concretist explicates modality just in terms of big, lumpy physical things, worlds, which are just mereological sums of smaller physical objects--no modal primitives, Lewis claims. And we should agree that such a theoretical advance would be worth almost any price in superficial commonsensical plausibility.

I grant a difference between Actualism and Concretism, in that the latter offers, as the former does not, the *prospect* of a reduction of modality.[22] But in fact, I believe, Lewis' reduction claim is untenable, by reason of the following further Actualist/Concretist difference:

(X) An Ersatzer has no problem with *impossible* worlds. An impossible world is just--e.g.--a set of propositions (one which happens to be inconsistent). But a Concretist is hard put to imagine how a physical, flesh-and-blood world could have logically incompatible constituents. Lewis himself simply refuses to countenance impossible worlds (1986, p. 151). For him, "world" just means "*possible* world." Thus it is itself a modal primitive even though it is not spelled like one. Insofar as the Concretist needs or wants to rule out impossibilia by fiat, the Concretist is stuck with a modal primitive just as the

Ersatzer is, and Lewis does not after all win issue VIII. (But this matter has become controversial; I shall return to it in Chapter 4.)

The ruling out of impossible worlds is a serious liability in its own right. For semantics needs impossible worlds. Though standard modal logics may trade just in possible states of affairs, the semantics of conditionals must deal with counterlogicals or at least countermetaphysicals, and the semantics of indirect discourse must deal with confused assertions, and doxastic logic must deal with inconsistent beliefs. Moreover, any Meinongian argument for the existence (in any sense) of nonactual possible worlds has an equally cogent parallel argument for that of impossible worlds; at least, I can think of no direct argument for "nonexistents" that does not support impossibilia by parity of reasoning. Nor have I ever heard a cogent objection raised against impossible objects that does not hold with equal force against nonexistents of any sort.[23] Modal metaphysicians have generally conspired to duck this issue,[24] but it will eventually have to be faced, and I do not expect anyone to find a reason (independent of Concretism) for countenancing nonactual possible worlds but refusing to acknowledge impossible ones.

However, in the face of my skepticism, Lewis (1986, p. 7, fn 3) has given a direct argument against impossible worlds; so let us address it.

5. LEWIS' ARGUMENT AGAINST IMPOSSIBILIA

Lewis notes that impossibilists want to seal off their contradictory objects by interposing a "modifier" or operator between (so to speak) the utterer and the contradictory states of affairs in question. This interposing tactic works well for, e.g., Linguistic Ersatzers: "There are round squares, though of course they couldn't really exist" comes out as "*According to some (inconsistent) world-story*, there are round squares," i.e., "Some (inconsistent) world-story has 'There are round squares' as a member." (I believe this to be an important virtue of Linguistic Ersatzism.) But Lewis notes that for him, the relevant operator is a quantifier-restricting modifier only. To him, "In some world distinct from our own" is a *locative* adverbial, like "On some mountain far away." And such modifiers, unlike hyperintensional sentential operators, "pass through the truth-functional connectives." Now, suppose that some traveller tells of a marvellous distant mountain on which there are round squares, or on which otherwise both P and ~P. "On the mountain, P and ~P" entails both "On the mountain, P" and "On the mountain, ~P." The latter in turn entails "~(On the mountain, P)." So the traveller (unlike the Ersatzer who posits inconsistent *stories*) has herself uttered an only slightly veiled contradiction.

...there is no difference between a contradiction within the scope of the modifier and a plain contradiction that has the modifier within it. So to tell the alleged truth about the marvellously contradictory things that happen on the mountain is no different from contradicting yourself. But there is no subject matter, however marvellous, about which you can tell the truth by contradicting yourself.

So much for our deluded impossibilist.

I do not yet see this. Remember that both Meinong's inclusive quantifier and Lewis' single quantifier run well past the actual, through the bizarre and into the physically impossible. Meinong contends that it ranges even into the logically impossible. If Meinong is right, then there *is* a subject matter about which you can tell the truth by contradicting yourself, if to put a contradiction within the scope of a restricted quantifier is indeed to "contradict yourself"; Lewis' premise to the contrary begs the question (against the impossibilist, not just against the Ersatzer). According to the friend of impossibilia who otherwise accepts Lewis' ontology and vocabulary, there are round squares, though of course only in marvellous regions of logical (or illogical) space. "On the mountain, ~P" is felt to entail "~(On the mountain, P)" because we would ordinarily be thinking of an actual or at least possible mountain. But once the friend of impossibilia is onstage and a powerfully inclusive quantifier is in the script, that entailment is simply broken. Thus, it seems to me, Lewis' argument fails.

If so, what becomes of the issue of modal primitives? Lewis must give the impossibilist some new reason why (M) is supposed to be false. And once more, I do not see how he will be able to do that without prior appeal to the fact that "worlds" containing round squares are *impossible*.

We have seen that a fair bit hangs on the Actualist-*vs.*-Concretist decision. Correlatively, the dispute will itself be informed by the arguments involved in the various issues I have listed.

It is only recently that our five approaches to the metaphysics of modality, and their various versions and sub-versions, have become as visibly distinct as I hope I have brought out so far. The primary point to grasp, and my main theme in this chapter and the last, is that anyone who needs to traffic in "nonactual" possible and/or impossible worlds and who is concerned to use these notions in a philosophically self-conscious and responsible way must (eventually) choose between our alternative approaches and *face the consequences*. Logicians, who remain cautious in their explanatory claims, may put off the choice indefinitely; but it is time for philosophical semanticists and modal metaphysicians to get serious.

Suiting action to the word, in the next chapter I shall motivate a version of Ersatzism and exhibit some of its advantages.

NOTES

[1] For present-day Combinatorialist theories, see Cresswell (1972), Skyrms (1981), and Armstrong (1989). These will be discussed in Chapter 3.

[2] Carnap (1947/1956); Hintikka (1961).

[3] We must give Russell enormous credit at least for crowd control: With but a few short (and unfair) articles, he kept Meinong not only at bay but anathematized--an embarrassing joke--for *seventy years*. Who in the social history of our profession can claim as much?

[4] Yet no one should think the distinction unproblematic. Zalta (1983, 1988) presents a rich and complex modal ontology that is neither obviously Actualist nor obviously Concretist. I think that according to preference it can bear either an Actualist or a Concretist interpretation. But in any case Zalta rejects Meinong's doctrine of *Sosein*; for him, the golden mountain is not ontologically golden, or a mountain.

[5] I believe it was first articulated by Tom Richards (1975).

[6] I adopt Lewis' now traditional name for the world we all share.

[7] A clue to Routley's prescience: As a Meinongian Concretist himself, Routley had been working on the world-individuation problem; and he had considered a spatiotemporal criterion, in a paper by himself and Valerie Routley (now Plumwood) presented to the 1978 Australasian Association of Philosophy Conference. Very likely he discussed the issue with Lewis.

More recently I have been similarly reviled by van Inwagen (1986, p. 212, fn 9), but unfairly so, for it was not Lewis (1986) that I had read as a Meinongian in 1979. The same goes for Linsky and Zalta (1991).

Lewis has since explicitly criticized Routley's Meinongianism (Lewis [1990]).

[8] It should be noted that Lewis himself rejects the Leibnizian theology of God as a transmundane surveyor of and chooser among worlds. Lewis thinks of gods as immanent to the worlds they rule--Christianity is true at some worlds, Norse polytheism at others, etc. A nice paper remains to be written on the implications of this for philosophy of religion.

[9] As was pointed out to me by Dorit Bar-On.

[10] Lewis addresses a different version of this point (aimed at his own view rather than at the "designated world" view) in (1986, sec. 2.5).

[11] That was what was wrong with van Inwagen (1980). Several of the premises of the arguments he went on to deploy against Lewis in that article are very plausible when understood against the background of the Actualist picture but hard or impossible to defend against the Concretist, and Lewis has always been a Concretist, whether Meinongian or not. (Van Inwagen has explained in correspondence that he turned Lewis into an Actualist "to save him from himself.")

[12] See, e.g., Lewis (1986, pp. 78-79).

[13] A referee has suggested taking "actual" as a predicate applying at each world to that world itself. That is certainly the way for the Concretist to go. But there is still the difficulty of running the result through Lewis' counterfactual semantics. (I do not say that difficulty is insuperable; the contrast with Ersatzism is only that the difficulty exists and is still unresolved.)

[14] Lewis (1968). It is interesting that Lewis foresaw this near-consequence of what was to become his distinctive ontology before publishing the ontology itself.

Stalnaker (1986) notes that the Concretist argument is not the only motivation one might have for moving to Counterpart theory. There are at least two others available to Actualists; one is even compatible with Haecceitism.

[15] In a seminar discussion of this material, Barry Loewer and Michael Hand expressed confidence that they would be able to work out such a view. I do not know whether they have since tried.

[16] Perhaps this will not entirely do; sometimes we may mean, not the real physical world, but the Ersatz "possible world" corresponding to it. But the former could still be the default referent. (It should be noted that van Inwagen himself uses the phrase "actual world" to designate the Ersatz; "The actual world is the comprehensive possibility that is realized" (p. 420). I grant that whether we can ostensively pick out the Ersatz "actual world," as opposed to the (actual) *world*, depends on the mechanics of deferred ostension. But I think I may be forgiven for assuming that direct ostension of the one world would, when deferred to the realm of abstracta, most naturally shift to the Ersatz *corresponding to it*.)

[17] Armstrong (1978; 1983, especially Ch. 8); Armstrong views uninstantiated universals as a failure of Naturalism. For Lewis' criticism of the Ersatzer, see (1986, pp. 159-65, 189-91).

[18] See also Adams (1975) and Lycan and Shapiro (1986).

[19] I grant that the Ersatzer has an equally nasty problem specifying haecceities for *nonactuals*. One might think Haecceitism obviously false for nonactuals anyway; but I somewhat reluctantly believe that to be a mistake: See Adams (1981) and especially Chapter 6 below.

[20] Kripke (1980), pp. 49-53. See also Rescher (1975, Chs. 3 and 4).

[21] Peter van Inwagen has protested to me that Kripke's stipulative view is as available to Concretists as it is to Ersatzers, for Concretists have always been counterpart theorists and counterparthood can be stipulated too. That is a revolutionary suggestion. Is counterparthood the sort of relation that could ever--much less, often--be stipulated? It seems to me that if counterparthood could ever be stipulated it could always be stipulated, and if it could always be stipulated then all *de re* modal fact would be stipulative rather than fact (which would be fine with Quine). Perhaps van Inwagen or Lewis would want to attack the first of those two conditionals, on the ground that when and only when similarity relations underdetermine the choice of a counterpart (as they do in my chimpanzee example), stipulation can appropriately bridge the gap. But still the modal "facts" resulting from those limited stipulations would, at best, differ ontologically from ordinary *de re* modal facts, which seems intuitively wrong.

[22] I am not altogether sure that Actualism is hopeless in that respect; see p. 358 of Lycan and Shapiro (1986). But I shall concede the point for purposes of this book.

[23] Naylor (1986) and Yagisawa (1988) have made this point also (but see Sharlow [1988]; I shall argue it in more detail in Chapter 4. N.b., impossible worlds are not, of course, closed under deduction (if there are to be more than one of them). But that is hardly a surprise; they are impossible, after all.

For further defense of impossibilia see Routley and Routley (1973, p. 230; also Routley (1976, pp. 247-250).

It should be noted that Lewis does countenance impossible *objects* (1986, p. 211); they are "objects" that overlap two or more worlds, and hence do not exist at any world, since to exist at a world is to be a mereological part of that world. But by the same token, they are not worlds, or even objects in any well-behaved sense; and each must necessarily encompass a spatiotemporal dislocation. Also, contra Lewis, the interworld overlap cannot be a sufficient condition of impossibility, for such an object's separated "halves" need have no mutually incompatible properties (though Lewis might argue that an internally dislocated object, since it spans more than one world, cannot have any counterpart occupying a single world).

[24] But see Hintikka (1975); Routley (1980, p. 291); Parsons (1980); Rescher and Brandom (1980); and again Yagisawa (1988).

CHAPTER 3

ERSATZING FOR FUN AND PROFIT

The Ersatzer's task is to find some system of actual objects that is structurally analogous or isomorphic to a system of possible worlds and therefore can *serve as* or *go proxy for* worlds.

1. POTENTIAL WORLD-SURROGATES

At least six sorts of system come to mind:

Linguistic entities. Historically, the most popular "ersatz" worlds have been sentences or sets of sentences. Carnap's "state descriptions" functioned as possible worlds. Hintikka (1962) followed a similar practice, though he dropped it in later works. If a "world" is understood as being a set of sentences, then possibility may be understood as *consistency*, and "actuality" neatly reinterpreted as *truth*. (Note that the metalinguistic approach fits nicely with metalinguistic theories of the propositional attitudes and provides handy objects for the attitudes.)

Propositions. One might move to sets of language-independent propositions, i.e., sets of abstract (but actual) objects having sentencelike semantical properties. This approach is defended by Adams (1974) and nicely elaborated by Plantinga (1974).[1]

Properties. Castañeda (1974, 1989) and Parsons (1974, 1980) have offered ways of construing Meinong's Objects as sets of properties. Castañeda's Guise Theory achieves a sort of fusing of Meinong with Frege and is able to treat a number of semantical issues quite neatly without any appeal to genuinely nonexistent possibles. A drawback here is that the approach requires the introduction of several new primitives. (It should be noted that in the end Castañeda declines his own offer; though he has the resources to eschew nonactual entities, he freely chooses the Meinongian interpretation instead of sticking with sets of properties. Parsons [1980] does the same.)

Combinatorial constructs. Quine (1968) suggested, and Cresswell (1972, 1973) elaborated, the idea of taking "worlds" to be set-theoretic combinatorial rearrangements of the posited basic atoms of which our own world is composed. I shall explain and discuss this view in more detail shortly.

Mental items. An obvious but so far unattempted move would be to take mental entities of some sort as our "ersatz worlds." Rescher (1975) is tempted by this approach and points out a number of its advantages, but does not adopt it in the end (cf. pp. 216-217). (See also McGinn [1981])

Ways things might have been. Stalnaker (1976a) suggests taking "ways things might have been" as *sui generis* elements of our ontology; "ways things might have been" are actual abstract entities in their own right, not to be reduced to items of any more familiar kind. (Zalta [1983] defends a highly ramified and articulated version of this view, on the author's preferred interpretation of his formal apparatus.) Stalnaker accepts Lewis' indexical analysis of "actual," but regards it as metaphysically uninteresting, since there is only one *world* for entities to be actual "at."

I argued in Chapter 1 that we have good reason for continuing to regard the inclusive Meinongian quantifier as a standard objectual one. And Ersatzism as a program has aesthetic attractions not unlike those of the analogous explications of number theory in terms of set theory. But any particular choice of a system of "ersatz worlds" may turn out to face philosophical difficulties of its own. For example, Lewis (1973, p. 85) argues that the metalinguistic method of Carnap and Hintikka is circular, in that "consistency" of sentences cannot adequately be defined save in terms of possibility. In addition, more typically, a choice of a system of world-surrogates may be seen to be technically inadequate, in that the intended isomorphism between the system and the lattice of worlds for which it goes proxy falls short in some way that frustrates our purposes in using pseudo-Meinongian quantification in the first place. A mentalistic approach, for example, is daunted by the paucity of *actual* mental events: The entire history of the universe will contain only finitely many mental entities, and it is hard to see how these might be parlayed into a system of proxies for all the multiply uncountable sets of worlds that must be posited for purposes of modal logic.

To illustrate some of the philosophical and technical limitations to which particular Actualist programs are subject, I shall discuss the Quine-Cresswell Combinatorial approach in some detail. I shall then conclude by making a tentative prognosis and recommending an alternative.

2. COMBINATORIALISM

To feel the initial attraction of Combinatorialism, consider the familiar notion of a chess game. Many chess games have been played, but there are plenty of other chess games that never have actually been played, though they are permissible according to the rules of chess. Yet we are little tempted to assimilate unplayed chess games to the golden mountain or the present king of France. A *game* is quite naturally understood as being a sequence of moves. A *move* we may take to be a triple whose members are a chess piece, an initial square, and a destination square (castling would require a small refinement). *Pieces* and *squares* are types, which may be regarded as sets of tokens. A game is played when two people under appropriate circumstances make a series of physical motions using (physical) chessmen and a board, which series mimics the sequence of triples that is the game in question. Some allowable sequences are never in fact mimicked in this way, just as some grammatical sentences of English are never tokened. And there is nothing at all metaphysically mysterious about this, save perhaps to nominalists who have qualms about sets. Could we extrapolate our Ersatzist metaphysic of possible chess games to cover all "nonexistent" possibles, including "possible worlds"?

Suppose that there are some metaphysically basic elements out of which our universe is composed. Call them "atoms" (in the metaphysical rather than the chemical sense). Our world, we may say, consists of these atoms' being arranged in a certain fabulously complex way. The actual *arrangement* of the atoms could be taken to be, or to be represented by, a vastly complex set, built up out of nothing but atoms and sets as members. Now let us construe "other possible worlds" as *alternative* arrangements of our atoms which mirror the ways our world might have been just as the actual arrangement mirrors the world as it is. These alternative "arrangements" are sets too, of course, and sets are actual entities.

The "atoms" are the fundamental building blocks of our own world, whatever those may eventually be shown by science or philosophy to be. Hypothetically, we might take them to be little particles, or occupations of spacetime points (Quine's preliminary choice), or Berkeleyan ideas, or whatever. But as I shall now argue, our choice of "atoms" may ultimately affect the adequacy of our Combinatorialist explication of nonactual worlds.

3. PROBLEMS FOR COMBINATORIALISM

Lewis (1973) pointed out that certain choices of atoms (certain decisions as to which actual things we ought to count as being the fundamental building blocks of the universe) commit the Combinatorialist to strong modal theses which would better be left as open questions. Schematically: Any Combinatorialist will end up ruling out some apparently imaginable states of affairs as holding in no possible world, on the grounds that these states of affairs cannot be construed as being arrangements of that particular Combinatorialist's chosen atoms. This means that any Combinatorialist's choice of atoms places substantive constraints on what states of affairs are to count as possible states of affairs.

Lewis does not really elaborate this point, but we can give some trenchant examples of what he is talking about, and we can also work the point up into a principled argument against Combinatorialism which will be hard to resist. The examples:

(a) It would seem to be possible that the world should have contained either more or less fundamental stuff. It is easy to envision an arrangement involving fewer atoms, or even one which would serve as the null world (presumably the null set). But how might we construct an arrangement corresponding to an increase in the amount

ERSATZING FOR FUN AND PROFIT

of fundamental matter? (One might think of representing new "atoms" by artificial means, such as pairing existing atoms with real numbers. This would be an appropriately Actualistic strategy, but would constitute an abandoning of the *Combinatorial* approach, since these new artificial "atoms" would not be the very sorts of things--atoms in the strict sense--that physical things would be *physically made of* in the alternate world. To put the point slightly differently: our hypothesis is that there might exist *atoms* that do not already exist in this world; and atoms are not arrangements of atoms and so cannot be represented as alternative arrangements of atoms.) It seems, then, that any choice of a stock of atoms commits the Combinatorialist to the *necessary* nonexistence of any more atoms, since there will be no arrangement and hence no possible world in which there exists an atom that is not one of our prechosen stock; any such extraneous atoms would be nonactual, and so shunned by the Combinatorialist.

(b) In making remark (a), I was thinking of alternative physical worlds. But our world contains abstract objects too. "Other" worlds might contain fewer abstract objects (no problem yet), or more (same problem as in (a)), or possibly even strange abstract objects not found here in @. How could we hit upon an arrangement mirroring a world that is mathematically deviant in this way? Here, perhaps, it *might* help to resort to artificial means (akin to Wiener's construction of ordered pairs out of ordinary sets), since unlike atoms, abstract objects are not noncomposite by definition. But it is hard to imagine there being a way in which the abstract objects of this world might combine to form new composite abstract objects in another world, but in which they do *not* combine in this world.

(c) What about irreducibly spiritual objects? Very probably there are no ghosts, monads, or Cartesian egos in this world; but there could have been, at least if we are to take seriously the views of brilliant philosophers who have believed in them. What sort of arrangement of atoms and sets could mirror such a state of affairs?

(d) As Lewis points out, we want to leave open the possibility that our world could have operated according to an entirely different physics and even according to a radically different geometry. It follows that any Combinatorialist who chooses either Euclidian spacetime or Minkowskian spacetime in which to locate his/her atoms will be oversimplifying at best. How might we allow for basic structural changes through worlds? It would be hard to motivate any further variegation of our procedures for forming wilder and more bizarre arrangements of the atoms of *this* world.

(e) I should think that *nonatomistic* worlds are also possible, such as one which consists of an undifferentiated miasma of Pure Spirit. A nonatomistic world can hardly be regarded as an arrangement of @'s stock of atoms.

Notice that in virtue of limitations such as (a)-(e), Combinatorialism also constrains our accessibility relation in modal logic, and hence restricts the modal systems we may countenance.[2] For example, Combinatorial accessibility will have to be asymmetric. A world having fewer atoms than @ is accessible from @, but @ cannot be accessible from it, since @ contains atoms that it does not contain and which therefore are nonactual relative to it. We are forced to conclude that S5 is too strong a modal logic, since S5 is based on a system of worlds whose accessibility relation is symmetric.

The obvious response for the Combinatorialist to make to (a)-(e) is to bite the bullet and maintain that the apparent "possibilities" that I have imagined are simply *not* possible. But there are two crippling rejoinders to this. (i) When the Combinatorialist says "not possible," this must be meant *in the strongest conceivable sense*. More atoms, spiritual objects, different physics, and so on must be impossible not merely in the physical sense, and *not merely in the metaphysical sense*, but in at least the sense in which overt contradictions are not possible. This is because otherwise there would have to be *logically* possible *worlds* in which such things obtained even if there were no physically or metaphysically possible worlds of that sort. And this is completely counterintuitive. It may be "impossible" for there to be Cartesian egos in some very strong sense or other, but not in as strong a sense as that in which it is impossible for the number 3 to be both prime and not prime.

(ii) Even if we collapse the distinctions between grades of possibility on which I have just insisted, the Combinatorialist still cannot deny that bizarre situations of the kind we have been talking about are *believed in* by some people. Descartes, for example, believed that there were nonextended, irreducible egos. Now, one of the main functions of possible worlds is to provide semantics for propositional attitudes. E.g., *belief* is said to be a relation between a believer and a set of worlds (his doxastic alternatives). Thus, it should be true in each of Descartes' doxastic alternatives that there are nonextended egos--or else we shall have to rob the possible-worlds apparatus of much of its interest by finding a different semantics for propositional attitudes.[3]

A slightly more sophisticated Combinatorialist ploy is that which Max Cresswell has taken in conversation.[4] This is to *refuse to make* a choice of particular items to serve as our atoms, and instead just to insist that there *are* basic objects in this world that play the modal role of atoms even though we shall never know or be able to say what they are. We are to find out what the atoms of our world are like by examining what is required to generate adequate modal semantics, and in no other way. This might seem to flout cherished principles of scientific realism, in that Cresswell's unspecifiable atoms are as inaccessible to physics as they are to the ordinary person, but Cresswell (I think) would respond that modal semantics *is* a science, and we are required by *it* to posit his unknowable atoms even if physics has no need of them and cannot discover them.

Further criticisms loom: (iii) This Tractarian ploy is paradigmatically ad hoc. We posit the special, unknowable, subphysical atoms *solely* for the purpose of saving Combinatorialism from refutation by entrenched modal intuitions. Accordingly, we render Combinatorialism untestable. We also ignore the ineliminable *epistemic* possibility, given whatever objects are in fact the atoms of our world, that objects of that sort should have been composites and not atoms. (iv) The ploy gets by some of our previous objections only at the cost of substantive metaphysical commitments. Take (c). Cresswell would have us believe that bodies and Cartesian egos (in such worlds as contain both) are just different kinds of composites out of more basic metaphysical objects; he is committed to "neutral monism" as a theory of the mental. Neutral monism might be true of minds and brains in this world, and it might be true even of Cartesian egos, but I doubt it, since *Cartesian* egos are supposed to be fundamentally different in kind from bodies. More to the point, we should not want our *semantics* to prejudge this venerable and complex metaphysical question if we can help it. (v) Cresswell's ploy fails even to address some of our objections, particularly (a) and (e). So if it helps at all, it does not help much.

A final criticism, one which applies to a number of other versions of Ersatzism as well: In (a) and (b) above, I have mentioned the possibility of throwing together a system of set-theoretic objects that might *ape* the group of "nonactual" things or worlds we need, in the sense of being structurally isomorphic to that group of things (cf. again Wiener on ordered pairs). Suppose, to take another example, that our world contains only finitely many physical atoms out of which physical objects are made, but that we want to construct nonfinitely many physical variations on our world. We might easily allow for this by pairing existing atoms with numbers, or some such. The resulting set-theoretic objects might be quite arbitrarily chosen, and the objects themselves might well have nothing intuitively to do with the metaphysics of modality, even though they were carefully selected for their joint ability to play the desired Combinatorial role. But why should we suppose that real *possibility* in this world has anything to do with pairs of atoms and numbers?

Frege would have rejected this question, and so do many of the intensional logicians who follow him on the methodology of positing abstract entities in semantics. ("*I* don't care what we let the quantifiers range over; the 'worlds' could be my dog, the Eiffel Tower, the real numbers, my grandmother's tricycle, and Bertrand Russell--just so long as the mathematics ends up dumping the right sentences into the right barrels.") This brings up Plantinga's (1974, Ch. 7, sec. 4) distinction between "pure semantics" and "applied semantics." A "pure" semantics does not interpret its quantifiers at all, and so does not commit itself to any particular domain. Its only adequacy-condition is instrumentalistic, that it predict the truth values of complex sentences given the truth and satisfaction values of their parts and that it capture the right sorts of felt implications. In effect, a "pure" semantics serves to axiomatize the predicate "valid modal formula" and nothing more.

But any number of interpretations of such axiomatizations will be available, no one of them having any better claim to "correctness" than any other (compare now the alternative ways of reducing number theory to set theory). Plantinga points out that a "pure" semantics "does not give us a meaning for '□', or tell us under what condition a proposition is necessarily true, or what it is for an object to have a property essentially" (p. 127; notice, incidentally, that the three tasks Plantinga mentions are mutually distinct). If our semantical theory is to *explain what it is* about our world that makes an alethic statement true in it, we must assign specific domains to its quantifiers, thus making it into an "applied" semantics.[5] (And if we are Actualists, our domain must contain only actual objects.)

A serious qualification is needed here. Modal and intensional semantics have been employed in aid of many different sorts of jobs and projects, not just the few enterprises mentioned in the preceding paragraph. And many of these projects, though technically demanding, are not philosophically ambitious enough to require any very specific ontological choice of domains for the Meinongian quantifiers. A trivial and obvious example of a justifiably noncommittal appeal to "worlds" is the casual, everyday semantical computation we do in routinely checking the validity of complicated modal inferences in the process of carrying on philosophical arguments. Perhaps a less obvious example may be the use of "possible-worlds" semantics in obtaining consistency and completeness etc. results for modal systems.[6] Possibly we can even learn something about the nature of necessity from "pure" semantics alone, though to articulate exactly what sort of illumination or understanding we gain here is much more difficult than most logicians admit. Then what uses of the "possible-worlds" apparatus do require a serious assignment of specific sorts of ranges to the Meinongian quantifiers?

I offer a paradigm case. Suppose someone were to argue as follows:[7] "Metaphysicians have always struggled with the problem of counterfactuals, wondering what fact it is in the world that makes a counterfactual true. Some have posited special dispositional facts; others have opted for what Quine calls an irresponsible metaphysic of unactualized potentials; still others have attempted deflationary but palpably inadequate metalinguistic accounts and such. Now, I can explain why metaphysicians have had such trouble with counterfactuals and have had to make such desperate lunges. The problem is that the metaphysicans have radically misconceived the issue: they have sought truth-makers[8] for counterfactuals *in the* (i.e., *this*) *world*. A counterfactual's truth-maker is not in the single world at which the counterfactual is true, but is transmundane, involving lots of worlds distinct from this one all at once. Once we see this, we can make great progress on the metaphysical problem of counterfactuals."

This argument seems very compelling, and so do similar arguments concerning the metaphysics of possibility and necessity. But notice that its *mainspring* is the metaphysical claim that our world is not the only world and that other worlds and their histories figure just as crucially in determining the truth of a counterfactual "at" our world. This metaphysical claim, if it is genuinely to explain how earlier metaphysicians misconceived the problem of counterfactuals, must be understood in substance as well as in form--its proponent must be able to offer some account of what his "other worlds" are if it is to provide the kind of metaphysical illumination that is being claimed for it. (If we are Actualists, of course, our "transmundane" facts will be reduced in turn to facts of our world; so for the Actualist the foregoing account will not be literally correct, but will illuminate just in the way that a model does before it has been successfully demythologized.)

It is at this point that the arbitrariness of our choice among different methods of slapping together "arrangements" makes itself felt. No one way of pairing atoms or sets of atoms with real numbers is more intuitive or natural than all others, for example, and so none has a better claim than all others to tell us what fact it is that makes some counterfactual or alethic or doxastic statement true.

In fact, the arbitrariness goes a bit further down; we need not appeal to the indenumerability of worlds or to the need for bizarre worlds to be troubled by it. Take Quine's way of constructing Ersatz worlds, in terms of occupations of space-time points. If a point is to be represented by a quadruple of numbers, what kind of set shall we choose to represent the occupation of a point? We might simply form the set of all the points that are occupied in a "world" and let that set be the "arrangement" that mirrors that "world." Or we might pair our quadruples with 1 and with 0 alternatively, representing occupation or vacancy respectively. Or the other way around. Or we might use the letters "O" and "V," as is suggested by the examples of restrooms on passenger trains. The alternatives are limitless.[9]

In light of this it is hard to see how a Combinatorial interpretation of modal semantics that has helped itself to artificial aids in this way could solve the problem of truth-makers even if it were technically adequate to lesser, merely computational tasks. But a proposal is made by Armstrong (1989) for doing away with the artificial aids.

4. TRACTARIAN COMBINATORIALISM

Armstrong has always been concerned to defend Naturalism in a very strong sense of that term, roughly the view that everything that exists (a) is located in (our) physical spacetime and (b) makes some causal contribution to the spatiotemporal world.[10] Thus in approaching modality as a Combinatorialist, he faces an obstacle right at the outset: Most Combinatorialists help themselves to set theory, because in order to fund the notion of "a rearrangement of" the basic elements of this world, some abstract individuating and ordering device seems clearly required. That means trouble for Armstrong, since a Naturalist is not entitled to nonphysical items like sets, at least so long as those items remain acausal and nonspatiotemporal.[11]

But following Skyrms (1981), Armstrong gives the Combinatorialist idea a bold Tractarian twist, endorsing the claim that the existence of this possibility or that one is determined by "the mere existence of" the basic elements involved (p. 37; cf. *Tractatus* 3.4). If that claim is really true, then perhaps Armstrong need not actually produce any set-theoretic or other abstract-object mockup in order to see "other possible worlds" as rearrangements of the metaphysical atoms of this world. (Further, he announces that his version of Combinatorialism will display a "fictionalist element," of which more shortly.) Let us see how this goes.

Armstrong begins with his already (1978) defended Tractarian ontology of states of affairs, maintaining that both "properties" and "individuals" are only aspects of and abstractions from states of affairs. States of affairs are causal and spatiotemporal, so Naturalism is maintained; Naturalism would be violated only if Armstrong were to permit uninstantiated properties, which he notoriously does not (1983, pp.). The world's ultimate building blocks are the atomic states of affairs, those consisting of a simple individual's having a simple property. As for Wittgenstein, every atomic state of affairs is logically independent of all the others (p. 41), and in particular, all simple properties and relations are compossible (p. 49).

Now, to the issue: What is a *merely possible* state of affairs? That, as Armstrong says (p. 45), requires a revision of his previous use of the phrase "state of affairs," since up till now he has used it to mean an actual element of the actual world. He begins with the notion of a false atomic statement. A merely possible state of affairs is what a false atomic statement purports to describe, and is ostensibly-referred-to by a gerund phrase, as in "a's being G." A nonactual *world* is a conjunctive aggregate of possible states of affairs. Armstrong begins with what he calls "Wittgenstein worlds" (p. 48), roughly those conjunctive states of affairs that involve all and only the actual basic individuals and properties of our world. In subsequent chapters he provides for "contracted" worlds that contain fewer individuals or properties, and for expanded worlds containing more individuals, though he argues against "alien universals." (There are many other refinements and elaborations as well, but I think none that will affect my assessment of Armstrong's basic position on mere possibilia.)

What is *combinatorial* about all this is (I presume) that the terms occurring in even a false atomic statement must denote--the subject must denote an actual simple, and the predicate must denote a universal that is actually instantiated by something. Thus if a whole merely possible world is a maximal heap of atomic states of affairs, the heap will be composed of actual individuals and actually instantiated universals merely rearranged, as is the Combinatorial *and* the Naturalist way. Again, *heap* or conjunction, rather than set or any other assembled abstract entity.

At this point there is a swerve: Armstrong disclaims the Ersatzer's project. Unlike all other Combinatorialists, he is not proffering a stock of actual entities to simulate or go proxy for possibilia. Rather, he "treat[s]...mere possibilities as non-existents" (p. 46).

A merely possible state of affairs does not exist, subsist or have any sort of being. It is no addition to our ontology. It is 'what is not'. It would not even be right to say that we can *refer* to it, at any rate if reference is taken to be a relation.

"Reference to" mere possibilia has the same linguistic, metaphysical and presumably epistemological status as does ostensible reference to ideal entities in science (ideal gases, frictionless planes, perfect vacuums); at least, Armstrong calls such ideal items "[t]he parallel" by which he officially explains the ontological status of his possibilia.

That is the fictionalist part. It is important to see that Armstrong's fictionalism is no afterthought, but is vital to his program: A given possibility, an actual or merely possible state of affairs, is a particular, Armstrong argues (p. 52); it is not repeatable within a single world. And the gerund phrase that designates or ostensibly-refers-to a state of affairs is a singular term, albeit a nominalized sentence. Thus, although there may be some sense in which a particular possibility is determined by "the mere existence of" the basic elements that figure in it, the possibility is not exhausted by those elements' existing separately or even by their Leonard-Goodman sum. It is rather, according to Armstrong, a unified particular from which the elements are merely "abstracted." If we were formalizing Armstrong's ontology, we would need a gerund operator, a singular-term-forming functor applying to name-predicate pairs; a *'s being G* would have to be expressed as something like $B\{a,G\}$.

Armstrong seems to grant all that.

When we talk about possibilities, we are talking about something represent*ed*, not a representation. (An ideal gas is not a representation.) (p. 46)

A possibility is a *thing* represented, but it is not on its face an actual thing. Thus it might seem that Armstrong must Ersatz if he is to save his Actualism; but unhappily, to Ersatz would forfeit Naturalism. The only way to avoid Ersatzing consistently with Actualism and the fact that possibilities are particulars is to insist that the mere possibilities simply do not exist and hence that descriptions of them are fictional. Thus one might say that Armstrong has destroyed his Combinatorialism in order to save it. (But since the resulting view is a piquant alloy, I shall digress a bit to examine it in the Appendix to this chapter.)

    The overall case against Combinatorialism seems serious. I have discussed it at such length partly just to illustrate the kinds of difficulties that an Actualist can incur (for professed Actualists often fail to notice these difficulties) and partly to urge Combinatorialists to switch to some less troubled form of Actualism. But to which one?

5. PROGNOSIS

    I believe that the only promising choice of actual entities to serve as "worlds" is that of sets of intensional objects. Thus, I would fall in either with Adams and Plantinga and construct worlds directly out of propositions, or with Castañeda and Parsons and construct possible objects out of properties and then group the objects into worlds on the basis of their stipulated interrelations. My main reason for choosing familiar intensional entities is that Actualist programs based on them do not seem to run into as serious difficulties of the kinds I have mentioned as do other Actualist programs. I think this is largely because they are posits of *semantics* to begin with: so there are (or seem to be) enough of them; actuality can again be explained in terms of truth; set abstraction on worlds remains ordinary set abstraction; there is no arbitrariness problem of the sort raised in section 3, since properties and propositions are characteristically introduced as being the meanings of predicates and sentences in the first place; and for the same reason the connection between "possible worlds" and semantical notions and alethic notions becomes quite straightforward and intuitive.

It is fruitful to allow real individuals to be constituents of propositions; a simple atomic proposition might be said to consist of a paired object and property. This policy provides an obvious basis for transworld identity and hence for quantifying in. It also obviates the "telescope" problem faced by the Concretist, which I shall expound further in the next chapter.

A very significant virtue of an Actualist account based on sets of propositions is that it makes room for *impossible* worlds as well as possible ones. As I argued in Chapter 2, we have just as great a theoretical need for Meinongian quantification over impossible objects and worlds as we do for admitting possible but nonactual ones. The propositional account also betters Combinatorialism by explaining something that would otherwise be paradoxical,[12] viz., how a self-contradictory proposition can be said to be *true*, even "at" an impossible world: The account rejects the Concretist's practice of analyzing *truth* as "truth at" our world. Rather, it analyzes "truth at" a nonactual "world" simply as set membership. The self-contradictory proposition is a member of the set of propositions constituting the impossible "world" in question; hence it is "true at" it.

Stewart Shapiro and I have formalized an Ersatz system of this kind (Lycan and Shapiro [1986]). I have not included that material in this book, because Shapiro did all the work and should get all the credit; but anyone concerned to see the details of the construction should consult that paper. I shall describe the system a bit further in Chapter 6.

A few objections need to be dealt with. First: Does the propositional account not run into the same problem of "too much necessity" that I raised for the Combinatorialist? For if we construct our "worlds" out of properties and propositions, how could we then represent nominalistic worlds containing no properties or propositions?[13] Reply: A world *in which* there are no properties or propositions, though, like any "nonactual world," is largely *made of* properties or propositions, is just a set of propositions one member of which is the proposition that there are no properties or propositions. A Carnapian state description theorist would likewise represent a world containing no sentences or other linguistic entities as a set of sentences containing the sentence, "There are no sentences or other linguistic entities." I can write a story about people who never write stories; there is no inconsistency here.

Second objection, suggested by Stalnaker (1976a) and vigorously prosecuted by Lewis (1986): The propositional account takes *mutual compatibility* of propositions as primitive, thus forfeiting the crucial benefit of explaining that notion in terms of worlds, and condemning us to explain the modal only in terms of the modal. Where is the intellectual advance? Reply: Though far from ideal, my being stuck with a modal primitive is no defect. What makes some propositions compatible or incompatible with others is no more mysterious, ultimately, than what makes some worlds possible and others not, or than what makes some stories describe segments of "worlds" and others not. I suspect any metaphysics of possibility is going to have to take some one of these easily interdefinable notions as primitive, and I see no lasting conceptual superiority in any one of the three approaches over the other two on this point. (Lewis himself *claims* to get by without any modal primitive; that claim will be disputed in Chapter 4.) In the meanwhile, Ersatz worlds still have their uses, for all the purposes listed on p. 1 of this book.

Third objection: What difference is there, in the end, between a *false proposition* and a *nonexistent state of affairs*? I have claimed the preferability of quantifying over the former to "quantifying over" the latter, but is the difference not merely terminological? Are nonexistent states of affairs not just as familiar as false propositions, belying my crucial argument in section 3 of Chapter 1?[14] Reply: (i) That crucial argument shows precisely why false propositions are preferable to "nonexistent" states of affairs. "There exist states of affairs that do not exist" has to be disambiguated, and my invocation of *falsity* achieves this, assuming we have an intuitive handle on what it is for a proposition to be false (or for a state of affairs not to *obtain*) that is prior to, or at least independent of, considerations of modal metaphysics. (ii) We do have such an intuitive handle, plus substantial grounds for expecting that it may be reinforced in a number of technical ways. For *falsity* is a familiar semantical property; and propositions (as they have traditionally been conceived) have *structures* relevant to the determination of their truth values. For example, a simple atomic proposition would be false iff its individual constituent did not instantiate its property constituent. Falsity for more complex propositions might be explained in terms of a more elaborate picturing or mirroring relation,[15] or it might be recursively defined á la Tarski. (In either case, care would have to be taken to avoid the semantical paradoxes and other possible cardinality problems.) (iii) The question of whether a proposition *exists* is a substantive and interesting question quite independent of whether the proposition is *false*. A nominalist might admit that the sentence "Russell is a genius" expresses a true proposition rather than a false one *if* there *exists* any proposition at all for it to express; and we may debate whether (e.g.) the proposition that quadruplicity drinks procrastination exists without doubting that if it does exist it is false.

As usual, of course, we shall have to take the rough with the smooth. There is a further objection to the propositional account that does reveal a serious failing: The account sacrifices the wonderfully elegant practice of explicating properties, propositions, and other Fregean intensions as being functions from possible worlds to extensions. If we reduce "possible worlds" to sets of propositions, for example, we cannot then reduce propositions to functions from possible worlds to truth values. This is an unpleasant price to pay, and both Concretism and Combinatorialism have the advantage here, but I think it is a small price when compared to the drawbacks that afflict those other theories of modality.[16]

APPENDIX: POSSIBILIA AS FICTIONS

Let us pause to survey prospects for Armstrong's fictionalism.

Granted, once we understand the semantics and the ontology of ideal entities, we might apply them to our nonactual "worlds." But now, what of the semantics and the ontology of fiction?

Armstrong himself says that what we need is "an Actualist, one-world, account" of fictional statements (pp. 49-50). (Evidently he is assuming that statements about ideal entities in science *are* literally fictional and so would yield to a general semantics of fiction. This assumption is plausible, but notice that we need not accept it, some other treatment of ideal entities might be preferable, and then we would have the choice of which treatment to extend to the merely possible worlds.)

Ideal scientific entities have a relevantly noteworthy feature: There is some inclination to say that statements about them are literally true, not literally false though true-in-fiction. Armstrong shares that inclination.[17] It could be gratified if we were to understood the relevant statements counterfactually, viz., as prefaced by "If there were any..." ("If there were any perfect vacuum, light would travel through it [in such-and-such a manner]"). But to extend *that* treatment of ideal entities to nonactual worlds would be to revert to the tactic of paraphrasing statements about possibilia in counterfactual terms. "There might have been a golden mountain" would be explicated in the usual way as "Some possible world contains a golden mountain," which in turn would be understood as fictionally asserting the existence of maximal conjunctive states of affairs that realize, *inter alia*, a mountain made of gold. Since (I presume) that fiction would be a true one by Armstrong's lights,[18] the counterfactual treatment of ideal entities would paraphrase it as something like (G) "If there were any array of possible worlds (i.e., any array of maximal conjunctive states of affairs besides the actual one that fit the Leibnizian picture of logical space), there would be one that realized a golden mountain."

(G) is hard to process and I think genuinely hard to understand. (Does the antecedent express the antiNaturalistic and perhaps Meinongian supposition that there are after all many *worlds* besides our own actual one?) But even if we understand what (G) says, we will have further trouble computing its truth-value; I do not know how to tell what would be the case "if Meinongianism were correct," though perhaps the Meinongian *picture* is clear enough to warrant the inference that there would be a nonactual world containing a golden mountain.

In any case, we have already seen deeper objections to explicating ordinary modal statements in counterfactual terms (sec. of Chapter 1), and good reasons in particular against paraphrasing talk of possible worlds in terms of unexplicated counterfactuals. Thus, although the counterfactual view of truths about ideal entities may be fine for Armstrong's standard scientific examples, it will not do if carried over to possibilia.[19]

Some philosophers have resisted the inclination aforementioned and held that as things are, physics is just plain *false* so far as it does commit itself to ideal entities. (They may add that reference to ideal entities is in principle eliminable from physics, so that physics can in principle be stated in such a way as to make it, and not just its empirical consequences, true; cf. Field's [1980] fictionalist theory of numbers.) That treatment would never do for Armstrong's purpose, since he wants everyday modal statements, at least, to come out literally true. He needs at least some sense in which possible-worlds statements are true, even though there is a more robust sense in which they are false.

Perhaps there are better ways than the counterfactual way of understanding truths stated in terms of ideal entities as literal, but none has come to me. Let us then, after all, stay with the idea of fictional truth. The obvious compromise (as between the counterfactual view of truths about ideal items and just rejecting the notion of truths about ideal items), is to say that a modal statement is literally true iff the corresponding possible-worlds statement is fictionally-true, even though the possible-worlds statement itself is not literally true. And my impression is that Armstrong would endorse that compromise.

We then have to give the Actualist semantics of fictional-truth itself. And as Armstrong plainly recognizes (p. 49), standard *possible-worlds* accounts of fictional truth are forbidden him. He must leave the mainstream and seek elsewhere.[20] Nor, of course, can he turn to Meinongian accounts such as Routley's (1980), Parsons' (1980) or Castañeda's (1989).[21]

What worldless fictional semantics shall he choose, then? Armstrong is offhand:

I do not know in detail what account to give, but it would be truly surprising if no such satisfying account were available. (p. 50)

And that, it seems, is that.

It is unclear why Armstrong thus shrugs off what seems clearly to be the biggest problem facing his theory of modality.[22] One cannot assume that a good theory of fiction (in particular) is just around the corner. For fiction is all too close to modality; fictional entities are nonactual possibilia, save for those which are impossibilia. The analysis of modality in terms of fiction is a step, but not a large step.

Let us therefore pursue a few avenues toward a non-possible-worlds semantics of fictional truth.

First, there are syntactic/inferential accounts. E.g., a statement S will be true-in a fiction F just in case S is a deductive consequence of some member of F (F being construed as a regimented set of sentences or formulas), where "deductive" means proof-theoretic rather than merely semantic.[23]

But to apply an account of this sort to Armstrong's "Wittgenstein worlds" would result in a specifically linguistic form of Ersatzism, whose mockups would be parochial to a particular formal language. First, it would require *an actual fiction* in which to ground Armstrong's ideal entities. Let us inaccurately but alliteratively call that fiction "Leibniz' Lie";[24] it says that alongside the actual world there exists an inconceivably huge panoply of other worlds, exhibiting structure R [of the sort needed to support intensional logic]. The structure R must somehow be spelt out, in detail, since all our subsequent statements about "worlds" must be strictly deducible from the Lie; and the Lie must be expressed in a particular language, say a formalized version of English, in order for deducibility to be defined on it. Then for a statement like "Some possible world contains a golden mountain" to be fictionally-true will be for there to be a formally correct deduction of the statement from the Lie.

This is not Ersatzism *in propria persona*, since it does not deliberately or directly furnish a set of actual world-simulacra. But it creates such a set indirectly: For any given world to "exist" for purposes of logic and semantics is (according to Armstrong) for that world to exist-in-fiction, but that is in turn for the statement of its existence to be actually deducible from Leibniz' Lie, and thus for that statement actually to exist. Thus, to every world there corresponds its existentially quantified specification-in-formalized-English, and if that correspondence holds, we might as well accept the specifications as Ersatz worlds and be done with it. But Armstrong does not want that. Moreover, a parochially linguistic Ersatzism faces well-known problems of its own.[25]

This last problem for the syntactic/inferential account actually expands into a general difficulty for any fictionalism regarding the Wittgenstein worlds, syntactic/inferential or not: Again, fictional-truth *tout court* requires a fiction. Statements are fictionally true or pretend-true only because there actually exist stories and other fictions for them to be true-in. But who authored a fiction according to which any one or more of the Wittgenstein worlds existed? It is not obvious that "Leibniz' Lie" actually exists.[26]

There is a further, though related difficulty. For "Some possible world contains a golden mountain" or any other specific possibilistic quantification to be deducible from Leibniz' Lie, the Lie must contain the appropriate stock of specific predicates, such as "golden" and "mountain." Thus the Lie cannot be a merely schematic description of a set of worlds, but must syntactically imply the existence (at some world) of everything that is in fact possible. But the Lie was supposed to be an *actual fiction*, historically tokened by someone in the real world. So for Armstrong's purposes (assuming he were to pursue the syntactic/inferential strategy) someone would have to have actually said something about gold and about mountains that entails the existence of a golden mountain. Philosophers using Meinong's famous example have done so, no doubt, but the same is not true for every single adjective-noun pair that in fact describes a possible object.

Armstrong's Combinatorialism might eventually rescue his syntactic/inferential incarnation from this objection, for someone--at least Armstrong himself--might actually say something about "all combinations" of predicates, along with a handy definition of a "combination" and some level-crossing principle to handle use-mention problems, and add all that to an original fiction that manages to contain all the primitive predicates; but something of the sort would have to be worked out at length, and we may be sure the details would be nasty.

The syntactic/inferential strategy is not hopeless, but neither is it very promising.

Secondly, there are local-ambiguity theories of fiction, according to which the words occurring in a fictionally-true statement have other than their literal meanings--at least, the copula in a subject-predicate sentence has a special meaning, which might be expressed as "fictionally-is":[27] It is not true that Sherlock Holmes lived in Baker Street (for the perfectly real Baker Street is not such that any Sherlock Holmes ever lived in it[28]); what is true is only that Holmes fictionally-did live in Baker Street.

I myself do not believe that there is any sense in which sentences like "Sherlock Holmes lived in Baker Street" are true save when they are used as abbreviating "In fiction F,..." statements. But it is clear that if I am wrong, and such a sentence *is* ever true at face value though in virtue of its copula's having taken a fictionalized sense, the sentence's nonexistent subject has turned Meinongian Object. If for "Sherlock Holmes lived in Baker Street" to be true is for Sherlock Holmes to fictionally-have lived in Baker Street, then Holmes is being treated as having ontological status of some sort. But for Armstrong, Holmes "does not exist, subsist or have any sort of being."

Also, were Armstrong to pursue the present line, the fictional copula would still itself need a semantics. What is the truth-condition of "Sherlock Holmes fictionally-did live in Baker Street"? If we are denied both possible worlds and any form of Meinongianism, I am at a loss.

Thirdly, there are Speech-Act accounts of fictional truth (e.g., Currie [1990]). On such a view, someone who asserts a fictional sentence meaning it fictionally is not making a literal declarative statement, but is engaging in a different sort of speech act and speaking with a different illocutionary force.[29] The felicity conditions on fictments (as we might call them) do not include literal truth; *contra* Plato, it is simply not a criticism of a work of fiction to point out that the work is not literally true. Of course fictments have many other kinds of felicity conditions, but not ones that are immediately relevant to modal metaphysics.

The trouble for Speech-Act accounts is that they still must distinguish between fictional truth and fictional falsity. The accounts differ in their means of explicating "fictionally-true" or "make-believe-true" and distinguishing fictionally-true from fictionally-false statements. But note that for Armstrong's purposes they will have to do that job without recourse to possible worlds and, if their proponents want them to differ significantly from the syntactic/inferential accounts aforementioned, fictional-truth cannot just be a matter of syntactic deducibility from a particular fiction either.

Speech-act theorists often sympathize with Gricean analyses of linguistic meaning and illocutionary force in terms of a speaker's (richly nested) propositional attitudes, and that strategy seems especially appropriate to the explication of fictive acts. Perhaps the difference between fictional truth and fictional falsity has to do with speakers' intentions and beliefs; though often the real world must cooperate in some ways if a statement is to be fictionally-true, far less is demanded of the world by fictional-truth than by literal truth. But here again, it is hard to see how the analysis would go without recourse to one or more of the means already denied to Armstrong--possible worlds, Meinongian Objects, a syntactic/inferential approach to the "In fiction F" operator, etc.[30]

I do not know what further alternative approaches there may be to fictional truth. But in any case it seems clear that Armstrong has most of the interesting work still ahead of him, even if we agree that "it would be truly surprising if no...satisfying [Actualist, one-world] account were available."

NOTES

[1] Plantinga actually talks of states of affairs which obtain or do not obtain and does not commit himself to identifying these respectively with true and false propositions (p. 45); I do not see that any fine distinctions here will be of importance to the metaphysics of modality.

[2] I owe this point to Phil Quinn.

[3] This point is somewhat weakened by the fact, already brought out and illuminatingly discussed by Stine (1973), that the failure of real people to be logical saints causes problems of this sort for possible-worlds semanticists in any case.

[4] I am indebted to Cresswell for a number of helpful discussions on this topic.

[5] I do *not* believe a parallel argument would succeed regarding number theory and set theory. I have proposed an analysis of numerical terms that blocks it, in Lycan and Pappas (1976, pp. 109-110).

[6] Thus, in no way am I suggesting that modal semanticists should interrupt their work and wait for our ontological problems to be solved to everyone's satisfaction. I do think that linguistic semanticists, who base their theories of *natural* language on intensional formal systems, need to be very careful in advertising the type of illumination that their theories yield.

[7] Something like this is suggested, though not expressly claimed, by Stalnaker (1968).

[8] I borrow this term from David Armstrong. I believe it was originally C.B. Martin's.

[9] Quine himself would reject this criticism and would reject the spirit of Plantinga's "pure"/"applied" distinction. In (1969a) he argues in effect that any choice of an interpretation for *any* quantifier is arbitrary in this way, or at least that it is relative to an arbitrarily prechosen *scheme* of interpretation. Thus, he finds Plantinga's and my demand inherently unreasonable. The important thing to see, though, is that if they are successful, Quine's arguments for this indeterminacy of ontology also impugn the status of semantics as a science and the whole idea of giving determinate "truth-conditions" for modal sentences or any other sentences, not to mention the whole idea of any sentence's having a determinate "meaning." Quine is radically skeptical of all such enterprises (on this, see Harman [1967, 1968] and Lycan [1984, Ch. 9]). So appeal to Quine's doctrine of ontological relativity will not help the Combinatorialist defend his own proposal for an applied semantics. Ironically, Quine himself (1969b, pp. 149-152) expends

a fair bit of energy in trying to remove arbitrariness of this sort which infects spatial and temporal dimensions, apparently in an effort to get at the *real* (nonarbitrary) members of his domain of arrangements. This seems quite uncharacteristic of Quine and remains for me an exegetical puzzle. A further mystery is that Quine seemingly invokes his "ersatz worlds" only to model the propositional attitudes of nonverbal creatures and does not seem to notice that his apparatus potentially generates a stock of arrangements adequate to deal with modal semantics generally.

[10] Armstrong's exact definition of "Naturalism" has varied slightly from work to work. His *defense* of Naturalism consists largely in his well-known Causal Argument, whose most recent version is presented in Armstrong (1989a, Ch. 1); see also Armstrong (1978, Ch. 12).

[11] Later, in Chapter 9, he sketches a program for Naturalizing set theory. (And see Armstrong [1991].) But he has an additional reason for eschewing sets in prosecuting the present project:

It seems that sets are supervenient on their members, that is, ultimately, things which are not sets. Supervenience, however, is a notion to be defined in terms of possible worlds, and hence in terms of possibility. It seems undesirable, therefore, to make use of sets in defining possibility. (p. 47)

[12] The problem was put to me by the late Don Mannison.

[13] This objection was raised by Tom Richards.

[14] This point was raised by Phil Quinn; Stalnaker (1976a) also hints at it, but instead argues rather neatly that his Actualist world-surrogates, "ways things might have been," are just as efficacious as sets of propositions and have a slight edge in economy. I think his argument can be resisted, but I would be almost as happy with his candidates as I would be with sets of propositions.

[15] I have suggested this in Lycan (1974).

[16] Besides, wonderful elegance or not, the reduction of intensional entities to functions on worlds has a philosophical drawback of its own, viz., it deprives us of our most obvious and natural answer to questions of the form, "In virtue of what is a in the extension of P at w?," that answer being, "In virtue of having, at w, the property expressed by P." Of course, some philosophers may have no care for such questions.

[17] "Some statements about ideal gasses, frictionless planes and economic men are true, while others are false" (p. 50).

[18] Which is not, of course, to say that there does actually exist a golden mountain, but only that "Some possible world contains a golden mountain" is one of the statements that would be accepted rather than rejected by modal metaphysicians, in the same sense as some statements about ideal gases are accepted rather than rejected by physicists.

[19] Notice two further points about ideal entities. First: Whether or not we think of truths about ideal entities as being *paraphrasable by* counterfactuals, we may agree that scientific talk of ideal entities *licenses* the same counterfactuals. Thus, any obscurity or weirdness in counterfactuals like (G) embarrasses Armstrong's analogy.

Second: Scientific idealization is often justified on the grounds that empirical consequences are unaffected. That too is a counterfactual matter; the idea is that empirical predictions are just (or approximately) as they would be if the idealizations held in fact, and that is why the idealizations are permitted. On Armstrong's analogy, then, it should turn out that empirical predictions are just (or approximately) as they would be if there really were worlds other than our own. But what sense could Armstrong, or we, make of that claim?)

[20] He adds that in any case "the notion of the fictional...has no special link with possibility," for there are impossible fictions (p. 49). Agreed: No *special* link, in that sense. But certainly a very important link, in that fictional entities are nonactual objects of intelligible discourse.

[21] Before leaving the topic of ideal entities, I want to mention a remaining troublesome feature of such entities as they figure in science: that typically they are degenerate or otherwise special cases of what would otherwise be real physical entities, where an important magnitude takes an impossible value, often 0). That is, they are limiting cases; and that is why we have so little trouble understanding the idealized language. (It also encourages the counterfactual analysis of ideal entities.) But Armstrong's merely possible worlds do not have that feature. They are not ideal limiting cases smoothly continuous with real physical objects. That too suggests that the analogy is not very helpful and that Armstrong would do better to stick by a fictionalist account less strictly tied to the foibles of ideal scientific items.

[22] I omit here a psychobiographical conjecture that was included in the article on which this section is based.

[23] Any plausible theory of fictional truth will expand this strict deductive-closure criterion by also including further sentences deduced or induced by the reader with the aid of real-world information brought to bear on his/her interpretation of the text. See Lewis (1978) and Ross (1987) (but for present

purposes try to ignore and abstract away from their having couched their accounts in terms of possible worlds).

[24] After having (I thought) coined that term, I remembered that Routley (1980, p. 96) had used it first, though with an entirely different meaning.

[25] Armstrong might reply that those very problems (e.g., there being fewer than continuum-many sentences of any actual formal language) constitute a sufficient reason for *not* simply taking the specifications as world-surrogates, and so much the worse for my argument. But each problem translates straightforwardly into a reason against the original claim that for a world to exist is for a statement of its existence to be deducible from the Lie.

[26] It might be replied that although a fiction is needed, an author is not. Fictions are, perhaps, Platonic entities existing in their own right, and only sometimes stumbled upon by creative writers. In that sense, perhaps, there is a fiction corresponding to each of the Wittgenstein worlds.
But (a) as before, that would collapse Armstrong's alleged Combinatorialism into Adams' "world-story" form of Ersatzism, and (b) no Naturalist can appeal to Platonic entities, "fictions" or anything else. A fiction may be a universal rather than a particular, but for Armstrong the only universals are immanent Scotist universals, only formally abstracted from actual physical states of affairs (1978; see also Armstrong's 1989b taxonomy of theories of universals).

[27] It would not help to maintain just that the quantifiers have special meanings, for that was our original problem about mere possibilia.

It is of course possible to locate the fictional/literal ambiguity of a sentence somewhere other than in the sentence's copula. One might hold that the predicates involved are paronymous, or even that the component singular terms shift their reference. (Castañeda [1989, Ch. 11] contains a useful critical review of these options.) For that matter, one might suppose that the ambiguity consisted in the presence or absence of an underlying "story-" or "In fiction *F*" operator, though that would just bring us back to the problem of providing a semantics for that operator and hence to the options already surveyed.

[28] Even in its 19th-century 200 block, which was half a mile south of the renumbered present-day 200 section. I am sorry to see that the latter is now falsely marked as the fictional home of Holmes.

[29] It is natural to gloss this by calling the speaker's act a *pretense*, as did Searle (1979); interestingly, Armstrong himself gestures toward a pretense theory of fictional discourse (1971, p. 444). But Currie (1990, Ch. 2) argues convincingly against implementing the Speech-Act view in terms of pretense.

[30] Currie offers an ingenious and attractive theory based on the notion of a work's "fictional author," which author is itself a fictional construct connived up by author and reader. (Specifically, it is true in a fiction F that P just in case an "informed" reader might reasonably infer that F's fictional author believes that P (p. 80).) This does very well as a criterion, applicable to particular works of fiction, for separating propositions into fictionally-true and fictionally-false-or-neuter, but it does not help explicate fictional truth ontologically, at least not within Armstrong's Actualist and Naturalist boundary lines. For (i) the "fictional author" is still a mere possibilium, and (ii) so far as has been shown, the business of the (hypothetical) reader's *inferring* something about the fictional author's *beliefs* requires the idea of a determinate set of propositions operated on by inferential procedures--a return to some version of the syntactic/inferential account.

CHAPTER 4

AGAINST CONCRETISM

As we have seen, Concretism comes in two flavors, neither of them vanilla. The Relentless Meinongian simply refuses to explicate either the inclusive quantifier "$(\exists x)_M$" or the more restricted actuality operator, while Mature Lewis equates "$(\exists x)_M$" with our own perfectly serious use of "There exists" but then explicates "actual" in terms of spatiotemporal dislocation.

1. UNDERSTANDING, OR NOT UNDERSTANDING, RELENTLESS MEINONGIANISM

Now that we have distinguished Relentless Meinongianism from any softer or more tractable doctrine and self-consciously deprived ourselves of any tacit reparsings or reductive explications of quantification over nonexistent possibles, let us assess it.

I have to take my place among those who find *Relentlessly* (i.e., *genuinely* or *primitively*) Meinongian quantification simply unintelligible. However: In saying this, I am not using the term "unintelligible" in its sneering bad-Wittgensteinian sense; so far as I am able to introspect, I am not expressing any tendentious philosophical *qualm*. (For this reason, my use of the term may be irrevocably misleading.) I mean that I really cannot understand Relentlessly Meinongian quantification at all.[1]

Further, I hypothesize that most people who do profess to understand it are tacitly assuming that there *is* some paraphrastic program, some reinterpretation of the "Meinongian" quantifier, or possibly some actualist domain that will make sense of the quantifier. My evidence for this hypothesis is that apparent Meinongians of my acquaintance, upon being pressed and confronted with suitably uncomprehending grimaces, have often been surprised that anyone would react in this way, infer that we must be talking past each other, and fall back on crude attempts at paraphrase or reinterpretation.

Yet I would quail at attributing any tacit assumption of this kind to Routley, Parsons, or the other real live Meinongians who have written imposing and subtle works defending the view. And, methodologically, I am not sure what argumentative force my unintelligibility claim has even if it is true for most people. For in some years' time a Relentlessly Meinongian scheme such as Routley's or Castañeda's, fully elaborated, may possibly have become fundamental to our semantical thinking, in such a way that younger philosophers will be taught it and work confidently within it without experiencing any difficulty of the sort I have forthrightly confessed to above. It may be that even I myself, if I follow developments and keep pace with the new generation, may come to share their linguistic facility and cease to have my now obdurate feeling of incomprehension. Should this prediction come true, how should we describe what it is that will have happened to me, or to the Meinongian quantifier, in the meantime?

It might be said (following Frege on the meaning of his term "sense") that during the interim the Meinongian quantifier will have taken on an indispensable role in the mechanics of a well-entrenched theory and that I will have come to grasp this role; this is why I will understand the quantifier then even though I genuinely do not understand it now. But this explanation will not do. For I already do grasp the explanatory role of the Meinongian quantifier, in the sense Frege intended: I already see and appreciate the functioning of the quantifier and its variables in semantical theories of the sort we all would like to bring to bear on the nasty philosophical problems they are designed to solve. So it is not understanding of the Meinongian quantifier's explanatory role that I lack now but will gradually acquire. Nor may we suppose that what I lack is knowledge of the real referents of the Meinongian variables, on the model of "gene" talk prior to the identification of "genes" as parts of DNA molecules: unreduced "gene" talk was understood (I presume) as placeholding, in that "gene" did duty for "whatever physical things play such-and-such a causal role in the mechanics of heredity"; but to treat "possible-world" talk similarly would be to envision (in fact, to demand) a reduction of "worlds" to actual objects, and thus to forswear Relentless Meinongianism and opt for some unspecified form of Actualism instead. So if it is true that the Meinongian quantifier somehow will subtly acquire meaning for me within the foreseeable future, I cannot easily describe this hypothetical process or find a model or precedent for it.

In any case, suppose that the Relentless Meinongian continues to take the strong line suggested above, insisting that "$(\exists x)_M$" *is* primitive, and that he/she does understand it even if I do not, and that no truth serum or searching psychoanalysis would reveal any hidden explicative program that mediates this understanding. Even if we go along with this and try to regard the Meinongian view as a competing theory which I just do not understand very well (rather than as gibberish), we will find some theoretical drawbacks that I believe suffice to motivate our seeking some other approach to possibility.

2. SOME DISADVANTAGES OF RELENTLESS MEINONGIANISM

Drawback 1: I have already pointed that the Relentless Meinongian is stuck with two primitive quantifiers, while a practitioner of any of our other four approaches to modality deploys but one. This point is by no means decisive, but is not unimportant either.

(Someone might suppose the disadvantage to be offset by a lack of *modal* primitives: If the Meinongian can explicate all modality entirely in terms of Objects and the admittedly excrescent "$(\exists x)_M$," that total reduction would be as impressive or more so than is any modal metaphysics containing an unreduced lump of modality. The conditional claim is correct; but unlike Mature Lewis, the Meinongian exuberantly allows "$(\exists x)_M$" to range over *impossible* worlds, so a modal primitive is still required to define possibility and necessity.)

Drawback 2: Embracing the two primitives, the Relentless Meinongian leaves our impasse as it is and dismisses my Quinean hostility and incomprehension as slow-witted or perverse. This position may be tenable, but it does not illuminate, indeed refuses to illuminate, what I have argued is the fundamental problem about possibilia. It resolves our contradiction (A), in effect, by announcing in a dramatic tone that the contradiction has been resolved and refusing further comment. (Compare the Cartesian Dualist on the subject of mind-body interaction.) Perhaps this is as it should be; perhaps the Meinongian way of looking at things is on the whole a better way than any of the four more conciliatory ways, and perhaps I *am* being stupid and perverse in failing to see this and to understand the Meinongian quantifier. But it would be nice if this were not so and if there were some metaphysical theory of possibility that brought feelings of improved understanding to all concerned and did not instigate a wrenching cultural conflict.

Drawback 3: A Meinongian realism concerning worlds implies that other worlds contain individuals who are very different from me but who are in fact either identical with me or (as Lewis would have it) counterparts of me in their respective home worlds. And, since the Relentless Meinongian does not regard this consequence as metaphorical or as a *façon de parler*, he/she must suppose that there is some objective fact about these individuals in virtue of which they do bear this curiously intimate relation to me, that we can know there to be such a fact, and perhaps that we can come to know the fact itself by hypothetical inspection of the individuals and their properties, using an imaginary "telescope." It is this commitment on the realist's part that occasions all the traditional problems of transworld identity and their Lewisian counterparts concerning counterparts.[2]

These problems, Kripke (1971) and Rescher (1975) have argued persuasively, are pseudoproblems generated by the "telescope" view itself and not by anything in the actual nature of modality. It is possible for me to have been a purple chimpanzee with yellow spots; therefore, we are told, there is a nonactual world some inhabitant of which is a purple chimpanzee with yellow spots and is identical with (or is a counterpart of) me. Suppose this world contains many other chimpanzees of just the same type. Which one is, or is a counterpart of, me? The Relentless Meinongian must assume that there is a determinate, correct answer to this question rooted somewhere in transmundane reality. But, Kripke and Rescher point out, to accept this assumption is far less plausible than to say that worlds are something *we stipulate* in imagining ways in which *we* might have turned out differently and that one of the purple chimpanzees in the world I am imagining, to the exclusion of the other chimpanzees, is me or is my counterpart simply because I, who am doing the imagining, stipulate that it is and that the other chimpanzees are only its (my) casual acquaintances. So we should reject the "telescope" view, and therefore also the Meinongian realism that implies it.

Drawback 4: Richards (1975) and Rescher (1975) have raised a more general epistemological point: If Concretism holds, how is it possible for us to know anything about other worlds, given that they are all "out there" in logical space independently of our mental activity and that they are causally and spatiotemporally inaccessible to us?

[The Meinongian's[3]] truth-conditions are such that, for any given [modal] statement, it is impossible in general to determine whether they are met and hence whether the statement is true. There is, however, a certain measure of agreement between people about the truth-value of certain modal statements. Insofar as there is agreement one must assume that if it is not catechised into the populace without any understanding of any truth-conditions for statements, then there is some other account of truth-conditions for these modal statements, and these truth-conditions are such that we may with some degree of confidence determine whether or not they are met. [Richards (1975, p. 109)]

There are really two problems here: That of how anyone, notably modal metaphysicians, could come to know there to be other worlds at all, and that of how any ordinary person could know the truth of an ordinary modal statement if the statement's truth-condition involves other worlds.

In addressing the second of those problems, Richards seems to be assuming that I cannot know a statement S without *first* knowing a statement T_s which expresses S's truth-condition unless S has simply been "catechised" or drilled into me by rote. If "truth-condition" here means the fact in the world which ultimately makes S true, Richards' assumption is made doubtful by a number of standard cases of philosophical analysis. Contra "The Philosophy of Logical Atomism," the *ordo cognoscendi* need not coincide with the *ordo essendi*. But presumably Richards means "truth-condition" rather in the slightly different Davidsonian/Tarskian sense, in which a sentence's "truth-condition" is the core component of the sentence's locutionary meaning. I should think Meinongians would accept this; Lewis, at least, intends his possible-world analyses of modal sentences to give those sentences' respective truth-conditions in this Davidsonian sense or something relevantly like it (cf. Lewis [1972]). And it does seem we come to *know* a sentence in part by processing that sentence's Davidsonian truth-condition;[4] it is not so obvious that the *ordo cognoscendi* is independent of the *ordo veritatis*. So I am inclined to think Richards is right in challenging the Relentless Meinongian to provide an account of how humans can know things about "other" concrete worlds.

Someone might protest that just the same nasty epistemological question arises for numbers (Benacerraf [1973]),[5] and while it is indeed nasty, it has deterred few philosophers and fewer scientists and mathematicians from quite robust realism about numbers; it is a problem, but all theories face problems.

Not so fast. Numbers are abstract entities to begin with; no one is being a Concretist about them. But our present target, the Relentless Meinongian, is a Concretist about other possible worlds. If the other worlds are supposed to be physical worlds in just the sense @ and you and I are physical, then the epistemological question has more bite, for our epistemology of physical objects is far better settled than is any we may have for abstract entities.

Of course, the only statements about other worlds that we need to know for semantical purposes are *general* statements. But how (Richards asks on p. 110) do we know whether there is a world in which Saul Kripke is the son of Rudolf Carnap? We cannot tell by inspecting worlds. We must inquire independently, on the basis of some test. Richards suggests that whatever test we do use constitutes the *real* truth-condition of "Saul Kripke might have been the son of Rudolf Carnap." This is overhasty, since we ought not to suppose that verification-conditions are truth-conditions (that way lies Analytical

Behaviorism and its ilk). The point remains, though, that the Meinongian's knowledge of what possible worlds there are and of other general truths about worlds is posterior, not prior, to his/her knowledge of what things are possible and what things are impossible.

Routley (1980, p. 352) contends that contra Richards, we can inspect or "inspect" the nonactual:

> [K]nowledge of nonentities may be obtained by a range of cognitive procedures, e.g. perception, imagination[,] dreams, memory, inference. We encounter nonexistent objects all the time in our intellectual and imaginative activities and sometimes in perception.... (p. 352)

In his sec. 8.10, Routley goes on to grant, indeed insist, that a Meinongian's epistemology will differ dramatically from the current naturalistic sort of epistemology presupposed by Richards' and Rescher's objection. No doubt; and we must grant to Routley that Objects are thinkables first and foremost, directly present to the mind on those occasions when they are present to it. But those observations give us little help with the problem of epistemic access to whole concrete, physical *worlds* alien to ours.

Lewis (1986) has replied to Richards on his own behalf; I shall take up the reply when we consider Lewis' own views.

The drawbacks I have listed are neither individually nor jointly decisive against the Meinongian position. I do not believe that it is possible to *refute* that position, unless the enormous cardinalities involved in a system of possible worlds should trigger some ingenious diagonal argument that is beyond my mathematical expertise to devise. I do think that the intelligibility problem and the drawbacks together are serious enough to provide strong motivation for seeking some alternative modal metaphysics. And I shall be making further unfavorable comparisons below. But let us turn to Mature Lewis, not a moment too soon.

3. MATURE LEWIS' MODAL REALISM

Lewis' modal metaphysics has always been known for its distinctively extreme or radical air, and he has earned much popular fame as an iconoclast. Although he modestly calls his view just "modal realism" (1986, p. 2), he should have chosen a more flamboyant label. For one can perfectly well believe in the firm reality of modal properties and modal facts without acquiescing in the claim that nondenumerable hordes of nonactual but grittily physical entities and worlds *exist* just as we do.

In more detail, Lewis has defended each of the following theses:

(i) There are nonactual possibles and possible worlds, and "there are" here needs neither scare quotes nor the semantic tolerance I have tried to extend to the Meinongians; nonactual possibles and worlds exist, in exactly the same everyday sense as that in which our world and its denizens exist.[6]

(ii) Nonactual objects and worlds are of just the same respective kinds as are actual objects and the actual world. Nonactual tables are physical objects with physical uses; nonactual human beings are made of flesh and blood, just as you and I are. [This is not just Concretism, which the Meinongian shares, but Concrete Completism, since Lewis admits no "incomplete" items.]

(iii) Nonactual objects and worlds are not reducible to items of less controversial sorts; worlds distinct from ours are simply *worlds*.

(iv) Quantifiers range over not all the actual individuals that there are but all the nonactual ones that there are as well, unless their ranges are explicitly or tacitly restricted in context.

(v) Expressions which distinguish actual individuals from among all the possible individuals, such as "real" and "actual," are really relational expressions holding between individuals and worlds; an individual *i* is actual only "at" or with respect to some world *w*. When we, in this (our) world @, call some object "real" or "actual," these terms are abbreviations for the indexical "worldmate of *ours*"; every possible individual is real "at" the world it inhabits.

(vi) All individuals, actual or merely possible, are worldbound; there is no genuine identity across worlds. You and I are not world-lines, but merely have *counterparts* in other worlds who resemble us for certain purposes but are distinct individuals in their own right.

(Again, it is important to see that most of the foregoing claims are independent of one another and that the discriminating theorist might well accept some of them but disagree with Lewis over others.)

On first hearing the view as Lewis expressed it in the 1970s, philosophers tended to respond with delighted horror and loud forebodings of incoherence. Not even Meinong had dared to suggest that nonactual individuals *exist* in just the same way as you and I do, or that somewhere out in logical space there are flesh-and-blood counterparts of me who are leading admirable lives of their own, sharing all of my virtues and none of my faults, even though as we all know and Lewis must admit in his heart, such people *do not* exist. But that was before Lewis had made his brilliant stroke of resolving the central (A)-disambiguation puzzle by defining the narrow Actualist quantifier as a specific restriction of the inclusive one.

Theses (i)-(iii), recounting the flesh-and-bloodiness of Lewis' worlds, account for some of his other views that might seem odd on their face: If worlds are physical universes, like planets only bigger and more drastically separate from each other, then it is fairly plain why Lewis accepts (v)[7] and (vi)--and why (vii) possibilia are grouped naturalistically into worlds in the way described and the spatiotemporal criterion is invoked, (viii) Lewis admits no impossible objects or worlds (p. 151), and (ix) he finds worlds are disanalogous to times,

CHAPTER FOUR

particularly as regards persistence of individuals through them (pp. 202-204).

For Lewis, to be "nonactual" is just to inhabit a world distinct from this one, though that alien world physically blooms and buzzes just as ours does and is in no ontological way subordinate. Thus, great importance attaches to the boundaries between worlds.

How then are worlds to be individuated? As we saw in Chapter 2, Lewis answers that what separates our world from the others is simply spatiotemporal disconnection. He toughs out (pp. 72-3) the obvious objection I mentioned, that is seems possible for a single world to contain a pair of spatiotemporally disconnected proper parts-- by denying that possibility. He grants my second objection, the potential weirdness of actual and possible spacetimes, by conceding (pp. 74-78) that worldmatehood might be determined by only "*analogically* spatiotemporal" external relations. That brings us to nonphysical worlds of various kinds. Lewis considers spirit worlds (arguing that spirits are at least temporally related to each other, though time would have to be extricated from @'s spacetime), and the null world, which he simply rejects. But there are others he ignores, such as purely mathematical worlds. As I have said, his world-individuating policy is vexed.

Lewis's thesis is that of a modal plenum: "unrealized" possibilities are not really unrealized, but are only spatiotemporally or analogically-spatiotemporally cut off from us. Now let us revisit Lewis' donkey example. Lewis claims that the possibility of talking donkeys obtains in virtue of there *existing*, in the absolute, ordinary, everyday, and physical sense of "exist," at least one donkey that has the two extraordinary physical properties of talking and of being spatiotemporally or analogically-spatiotemporally disjoint from us.

Now as I hinted in Chapter 2, I think that this claim is grotesque; at least, I say, no one will ever have the slightest evidence for believing either that there exists a talking donkey or that there exists a donkey that is disjoint from us. If the target sentence "There might have been talking donkeys" is more or less obviously true, and its proposed analysans is indeed garishly false, then Lewis's theory is seriously embarrassed.

4. THREE FURTHER OBJECTIONS TO MATURE LEWIS

Let us return to the problem of knowledge originally raised by Richards and Rescher, for Lewis (1986) addresses it.

I argued that the problem is worse for the Relentless Meinongian than is Benacerraf's corresponding problem for the mathematical Platonist, precisely because the Meinongian's worlds are concreta. I can now argue that the problem is even worse for Mature Lewis than for the Meinongian. The point is well put by Skyrms (1976): "*If* possible worlds...are supposed to exist in as concrete and robust a sense as our own..., then they require the same sort of evidence for their existence as [do] other constituents of physical reality" (p. 326). But there is not the slightest physical evidence for the existence anywhere of donkeys that talk. The reason this should embarrass Mature Lewis more than it does the Meinongian is that Lewis claims the talking donkeys exist in exactly the same sense as that in which real donkeys do.

Lewis replies to Skyrms that "[o]ur knowledge can be divided into two quite different parts" (p. 111): Contingent knowledge requires causal contact with its objects, but "[o]ur necessary knowledge...requires no observation of our surroundings, because it is no part of our knowledge of which possible world is ours and which possible individuals are we" (p. 112). For the case of mathematical and modal knowledge, we make "imaginative experiments," and "thereby cover an infinite class of worlds all in one act of imagining" (p. 114). Lewis invokes a previously defended (though qualified) "principle of recombination," to the effect that "anything can coexist with anything else (p. 88); the idea seems to be that we imagine recombinations of actual things, and infer that such recombinations are possible, hence (?) exist at an alternative world.

I suspect that as a piece of psychology this conjecture is right. But I do not see that it, or the mental process it describes, provides anyone with *evidence* of the physical existence of talking donkeys spatiotemporally disconnected from us. (Nor does Lewis get to assume, in addressing the objection, that the physical existence of such donkeys *is* a *modal* matter, for that would squarely beg the question.) And so I do not see that Lewis has answered the objection.

Secondly (I merely swipe this point from van Inwagen [1986], Plantinga [1987] and Armstrong [1989]), suppose we *were* somehow to find out that Lewis' general analysans is true, i.e., that there do exist cardinalities upon cardinalities of other physical worlds, all spatiotemporally disconnected from us and from each other. (God might reveal this, or physicists at Princeton or Cal Tech might find that it resolves all the paradoxes of quantum mechanics.[8]) But then surely we would have discovered something about reality, that there really exist other physical spacetimes merely dislocated from ours.

[T]here is no reason to think that what is isolated from us is merely possible relative to us. On the contrary, if it exists, then it is real; it is actual. (Armstrong [1989], p. 16)

The spacetimes would be actual regions of reality, rather than merely possible "worlds."

Further, Plantinga argues, no one would think the discovery had anything to do with *modality* in particular, at all; it would not be taken as vindicating any thesis about possibility.

Surely I could have been barefoot even if everyone, even those in other maximal objects if there are any, were wearing shoes.... [W]hat is a possibility for me does not depend in this way upon the existence and character of other concrete objects....
Of course there is *something* in the neighborhood with respect to which Lewis is a realist, and a pretty unusual and interesting thing at that: a plurality of maximal objects.... Lewis is certainly a realist of an interesting kind; but what he isn't is a *modal* realist. (pp. 209, 213)[9]

(Though Lewis might fairly reply that this initial, untutored absence of intuitive connection has little argumentative weight as against the considered theoretical advantages of his theory.)

A third and final shot, originally fired in Lycan (1979):[10] Lewis' (ii) raises an awkward question. It follows from his view that just in virtue of the logical possibility of my having climbed Mt. Everest and proved Gödel's Theorem upon reaching the top, there is, somewhere out in logical space, a flesh-and-blood person who is just like me except for having accomplished this *tour de force*. Now, if this person and other various counterparts of mine in other worlds *are* all flesh-and-blood people who resemble me in certain important respects, it would be fascinating for me to be able to meet some of them personally and talk about the interesting things we have in common. What prevents me from travelling to another world and meeting my counterpart at that world and discussing life with him? Of course there cannot be a logical spaceship that allows me to traverse logical space in this way. But why not?

True, worlds are causally and spatiotemporally disjoint from each other and may even differ in their physical laws; this fact suffices to explain why my counterparts in other worlds are forever cut off from me and why my visiting them is impossible "in principle." But I would like to say that the idea of "visiting one of my counterparts," *so far as this idea is occasioned by the mere possibility of* my having climbed Mt. Everest, etc., is not a pleasure which is *denied* me by scientific or even metascientific obstacles, but is rather nonsense. The impossibility of crossing causal or spatiotemporal dislocations, admittedly a very high *grade* of impossibility (stronger than physical, though weaker than logical), fails to do justice to my more nihilistic intuition to the effect that the mere possibilities of everyday life do not or should not occasion even an intelligible scenario of the sort at issue.

Lewis (1986, p. 81) replies to the "logical spaceship" argument. "No trans-world causation, no trans-world causal continuity; no causal continuity, no survival; no survival, no travel." That is: The personal continuity required for the duration of intermundane travel in turn requires a sustaining causal chain, but the causal chain would of course have to cross world boundaries. According to Lewis' (1973a) counterfactual analysis of causation, trans-world causation would in turn require the truth of certain counterfactual statements about worlds, which statements turn out to be undefined in Lewis' (1973b) semantics for counterfactuals. If all this is correct, then my talk of "visiting one of my counterparts" ultimately is undefined, and this explains its perceived nonsensicalness. Lewis adds (1986, p. 80) that we tend fallaciously to "think of the totality of all possible worlds as if it were one grand world, and that starts us thinking that there are other ways the grand world might have been." (Cresswell [1990, p. 166] concurs that Lycan [1979] "seems to have [been] misled" in that way.)

My only objection to this line of reasoning is that, for independent reasons, I believe that *given Lewisian Concretism*, the counterfactuals in question ought not to be left undefined by an adequate semantics for counterfactuals,[11] and the "grand world" picture is not, or not obviously, misleading. Suppose there do exist, in addition to @, a host of other physical universes differing from @ only in detail, and spatiotemporally cut off from us; and never mind for the moment that those universes are supposed to have anything to do with modality or with the analysis of counterfactuals. Then it seems to me counterfactuals and even counternomologicals about those universes *ought* to make sense. Whyever not? They are physical, just like ours, and the only particularly odd thing about (many of) them from our point of view is their spatiotemporal dislocation from us. Why should the dislocation *alone* prevent us from framing interworld counterfactuals? And remember, the counterfactuals need only make sense; I am not suggesting that any of them is ever true.

(It may be thought that in saying "Never mind that the universes are supposed to have anything to do with modality or with the analysis of counterfactuals," I have begged the question against Lewis, for his already defended claim is of course precisely that the universes *are* the possible worlds in terms of which counterfactuals are to be analysed. But I am not simply assuming the falsity of his views on counterfactuals. I am (I hope) remaining neutral on them and taking no theoretical stance in regard to counterfactuals. Rather, I am starting with Lewis' Concretist metaphysic and arguing that if it is correct, then interworld counterfactuals should make sense, whatever analysis might later be given of them.)[12]

5. LEWIS AGAINST US ERSATZERS

I have suggested that Lewis's modal metaphysics is wildly false. But Lewis himself admits its drawbacks, and claims only that competing accounts of possibilia are equally unsatisfactory. Most specifically, he rejects all forms of Ersatzism.

He also admits that Ersatzist theories ostensibly accommodate the phenomena of modality, folding those phenomena into our overall account of the universe, wholly in terms of actual states of affairs, without reliance on any possibilia taken as primitive. He even grants that "the ersatzers have the advantage in agreement with common sense about what there is" (1986, p. 140), but still he calls Actualism an ill-advised grab at "paradise on the cheap" whose seekers in fact end up "pay[ing] a high price," and in his Chapter 3 he tries to defeat it.[13]

He distinguishes three forms of Ersatzism: "linguistic," "pictorial," and "magical"; but of these he respects only the first, which construes "possible worlds" as, or replaces them by, sets of (somehow) linguistic entities. Lewis offers two main objections to linguistic Ersatzism. The first is that any set of sentences or propositions purporting to describe a possible world must be a consistent set, and consistency is a matter of possible truth; for this and several other reasons (pp. 151-157) the linguist Ersatzer is stuck with a primitive modal notion and cannot claim to have explicated modality in non-modal terms.

The question of primitive modality is tricky. I know of no Actualist who has been able to dispense with it (my own current modal primitive is "compatible" as applied to pairs of properties). But, if every Actualist is stuck with some modal primitive, so, I would say, is Lewis. The flesh-and-bloodiness of his worlds might be thought to relieve him of the need for the sort of abstraction indulged in by Actualists, and so it does; but Lewis mobilizes a modal primitive nonetheless. It is "world." "World" for him has to mean "*possible* world," since the very flesh-and-bloodiness aforementioned disinclines him to admit impossibilia. Some sets of sentences describe "worlds" and some (the inconsistent ones as we know them to be) do not; but Lewis cannot make that distinction in any definite way without dragging in some modal primitive or other. Thus, Actualists are no worse off in that regard.

Lewis disputes this claim, so let us dwell on it some more. Here is a slightly different way of making the point: As before, "It is possible that there should have been talking donkeys even though there aren't really any" comes out simply as

(L) $(\exists x)(Dx \ \& \ Tx \ \& \ (n){\sim}(ST(x, \text{us}) = n))$,

where "ST" is read as "the spatiotemporal distance between." Not only does that formula contain no clearly modal terms; it does not mention *worlds* at all. So on the face of things, my argument that "world" is modal for Lewis does not even get started against the Mature theory.[14]

But wait. Let us stage a confrontation between Lewis and a misguided Meinongian who is championing round squares. The idea is to see how Lewis tries to rule the impossible objects out-of-bounds. Our Meinongian says, "Along with everything else, there are things that not only do not but could not exist, such as round squares." On Lewis' Mature analysis that utterance comes out as

(M) $(\exists x)(Rx \ \& \ Sx \ \& \ (n){\sim}(ST(x, \text{us}) = n))$.

(I shall assume throughout that "R" and "S" are mutually incompatible predicates.) Now, Lewis must come up with a way of showing that statement to be false, *without* saying anything like "No *possible* world has a thing like that true in it."

Granted, the Meinongian's impossibilist formula (M) is already a formal contradiction, conjoining "Rx" and "Sx." Is that not enough to show that (M) is false, without dragging in any modal notion, so long as we know that any formal contradiction is false?[15]

Shades of Russell: What could be better proof of absurd falsity than overt contradictoriness? But (shades of Meinong in reply:) contradictoriness is a fault only in actuals and in possibilia, not in admitted impossibilia. Of course there could not actually be round squares. But there are nonactual, impossible round squares, and so

Lewis ought to admit that they exist in worlds spatiotemporally cut off from our own.

Miller (1989) has defended Lewis against my charge by formulating an explicit Lewisian definition of "world," designed specifically to be nonmodal. He invokes the notion of spatiotemporal relatedness that Lewis uses to individuate worlds:

(1) Individuals are worldmates if they are spatiotemporally related. (2) A world is a mereological sum of worldmates.[16] (p. 477)

No modal terms there, Miller contends.

I have two objections to make in response. First, since nothing but the impossibility of round square cupolas in the first place keeps a round square cupola from being spatiotemporally related to another, perhaps less exotic object, Miller's Lewis would not be able to rule out *worlds* containing round square cupolas. The real Lewis would have to amend (2) to read, "A world is a mereological sum of *possible* worldmates [i.e., logically acceptable candidates for worldmateship]."

Second, the spatiotemporal relatedness must itself be understood as the bearing of a *possible* or logically acceptable spatiotemporal relation; for Lewis is forced as we have seen, by the evident possibility of spacetimes alternative to ours, to grant that worldmates may be only "analogically"-spatiotemporally related. Without that tacit understanding, any object including any given round square cupola *is* spatiotemporally related to the (actual) Sydney Harbour Bridge--albeit by some logically incoherent relation.

Thus Miller's explicans is modal, indeed doubly modal, after all.[17]

Lewis's second objection to Ersatzism is that the linguistic Ersatzer cannot distinguish possibilities that are, in fact, distinct possibilities. In particular (pp. 159-165), there might have been alien "natural properties," properties which do not actually exist but do adorn worlds having (e.g.) an alien physics. Since those properties do not exist here, we have no way to name them, and so we cannot describe worlds in which they do exist, worlds in which they are switched around, etc.

Lewis has in mind what he calls "sparse" properties (p. 60), the genuine physical properties which make for real similarity and which make a physical world go. It is easy to suppose that other worlds' fundamental particles might have unusual spins or "flavors" or the like, unknown in the actual world. But let us recall that, for Lewis, properties, sparse or not, are just sets of this-worldly and/or other-worldly individuals; such sets exist and have their members independently of what world one finds oneself in. Thus, the "alien" properties are only here-uninstantiated properties, i.e., for us, *uninstantiated properties*. There is nothing peculiar about that, at least for anyone who countenances properties in the first place.

Lewis anticipates the latter point and responds to it (pp. 160-1) by asking, "[W]hat does the Ersatzer mean by 'property'?"; an Ersatzer's "properties" are usually taken as primitive, since they are needed in the building of Ersatz worlds to begin with. One might observe that the notion of a property is intuitively more familiar than that of a possible world or that of a physical but nonactual object, and so it is less offensively taken as primitive. But uninstantiated properties are especially problematic, in that they cannot be abstracted from instances but are only "of a kind with" the better-understood properties that have been abstracted from actual instances. Lewis goes on (pp. 189-191) to complain that such properties can be supposed to "do their stuff" only by magic. But I do not see how much more serious a charge this is, than just that they are indeed taken as primitive.

6. SUMMING UP

In my view, the following box score emerges from our survey of these issues.

I have contended that Relentless Meinongianism is unintelligible and Mature Lewis is calamitously false. If true, those contentions ought to settle the matter of Concretism. I think they are true, but I can hardly claim to have *proved* that. In case they are not true, I have tried to take each Concretist view seriously and have made several substantive objections to each.

But finally, there are the ten issues surveyed in Chapter 2, which served there to distinguish Actualism from Concretism as sharply as possible. Here I shall argue that they lend some weight to the case against Concretism as well.

Actually the Concretist can claim victory regarding issues (IV), understanding the designator "this world," and (V), uninstantiated universals. And Lewis does in addition claim victory regarding issue (IX), reduction and modal primitives, though I remain unconvinced. But the Ersatzer clearly wins (I), distinctness of worlds, (II), counterfactuals about worlds, and (X), impossible worlds, which seems a fair trade for (IV), (V) and (IX). Further, I believe we should side with the Ersatzer in the admittedly less settled matters of (III), transworld identity *vs.* counterparts, (VI), the world/time analogy, (VII), Haecceitism, and (VIII), the stipulative *vs.* the "telescope" view. If so, a weight of evidence is shifting toward Actualism, even if we disregard Concretism's incomprehensibility or falsity *and* the objections made against the two Concretist views earlier in this chapter.

NOTES

[1] Lewis (1990) joins me in this. Or rather, he distinguishes two possible translation schemes for Routley's quantificational language into the one-quantifier version of English that he and I both believe we speak; what he finds unintelligible is Routley translated as he, Routley, means to be translated, and for this reason he advocates the other translation scheme, which would bring Routley's view far closer to (Mature) Lewis' own.

[2] See originally Chisholm (1967) and Feldman (1971).

[3] As his title hints, Richards' essay is a critique of Lewis in particular. But the exegetical issue of whether pre-Mature Lewis qualified as a Meinongian does not affect the present argument, which succeeds or fails against the Meinongian whether or not Lewis was one.

[4] I defend this claim in Lycan (1984).

[5] Lewis (1986, p. 108) says, "The objection echoes Benacerraf's famous dilemma for philosophy of mathematics."

[6] Lewis (1973, pp.) offered a direct positive argument for his view, besides the theoretical advantages he claims for it. I shall omit discussion of the argument here, since I have criticized it in Boër and Lycan (1975b), and since several other commentators and reviewers have criticized it as well, some more effectively than I. See Nute (1976), Richards (1975), Haack (1977), and Stalnaker (1976a).

[7] But see also pp. 93-4.

[8] Yeah, right.

[9] Recalling that Benardete (1964) considers the concept of a pluriverse as an example of a case in which we might need the higher infinite cardinals to count physical objects, van Inwagen reports that "[i]t never occurred to [Benardete] that it would be appropriate to describe parts of other universes in the pluriverse as mere *possibilia*, and he still doesn't see it" ([1986], p. 199). Thus also Armstrong:

C.D. Broad...argued that spatio-temporal systems entirely cut off from each other are a logical possibility.... I read Broad on the point, and thought he was right. I still do so. But I did not think of such 'island universes'...as anything but *actual* relative to our spatio-temporal system. Neither did Broad. ([1989], p. 16)

[10] But I seem to remember that this point was suggested to me by Steven Boër.

[11] On this point, see also Richards (1975, p. 108).

[12] Cresswell (1990, pp. 166-167) makes a different objection to the spaceship argument: that "Lycan's problem is not just for a possible worlds theorist but for anyone who takes modal talk seriously." If Cresswell can show that, then I will agree that my argument is bankrupt, since the argument is supposed to proceed from two very special features of Lewis' modal metaphysic (his existence claim and his bare spatiotemporal criterion of worldmatehood). But I do not yet understand Cresswell's ensuing discussion, though it is based on his claim (Part I) that any semantics adequate to the expression of certain natural-language constructions, even if it purports to leave its modal operators primitive, will have the power of quantification over worlds, willy-nilly.

[13] Lewis seems to identify Actualism with Ersatzism. But I would call Kit Fine a non-Ersatzing Actualist (see Fine and Prior [1977]), as is Zalta (1983, 1988) on one interpretation.

[14] Van Inwagen (1986) joins Lewis in claiming that, once we firmly distinguish the Mature view from Meinongian Concretism, Lewis needs no modal primitive at all for his *analysantia*.

[15] Does it matter *how* we know that? Suppose we argue that the only way to know it is by truth-table and that truth-table reasoning is just possible-worlds reasoning. Offhand, this seems right to me, though it is not the main issue.

[16] I assume Miller intended a *maximal* sum, but that question is irrelevant here.

[17] My dispute with Miller is sorted out much more deeply in Pigden and Entwisle (1991).

Recently both Lewis and Robert Stalnaker have explained to me in correspondence that they are profoundly unconvinced by my argument. I have not obtained permission to quote their counter-arguments here, but both seem to reject my assumption that it is tendentious, stipulative, or otherwise committal to ignore impossible worlds. A referee has chimed in on their side. But I am remorselessly unconvinced by their unconvincedness, and see dialectical subtleties here that will take some effort to sort out.

CHAPTER 5

ESSENCES

I have declared a preference for Haecceitism in regard to actual individuals, but as yet have done nothing to support it.

1. THE CASE FOR HAECCEITISM

Let us review the alternative approaches to matters of individual essence. I take there to be four possible sorts of answer to the question, "What makes me *me*?," where what is wanted is a set of individually necessary and jointly sufficient conditions:

(i) The view that individuals have qualitative essences, expressible by (more or less) Russellian definite descriptions, and that these essences are "accessible" individuating properties or sets of them, in that they are the sorts of everyday properties to which we have easy epistemic access. E.g., they are the sorts of properties by reference to which we actually do recognize and identify other people in practice. On this view, my own essence would be something like $\lambda x(x$ is the North Carolinian philosopher who wrote *Modality and Meaning*).

(ii) The view that individuals have qualitative but *in*accessible essences. These individuating properties too would be expressed by definite descriptions, but are arcane and scientific or at least not detectable by the general public. If Kripke is right, for example, my own essence would be something like $\lambda x(x$ is the outgrowth of such-and-such a particular sperm cell and egg), or λ*x*(*x* has such-and-such a determinate genetic code).

(iii) Haecceitism, the view that individuals have essences but not qualitative ones at all. On this view, my essence is simply λ*x*(*x* = WGL), the property of *being Lycan*, period.

(iv) The Quinean view that individuals do not have essences at all.

These options logically exhaust the field. Indeed, for that matter, the fourth seems ruled out *a priori* by the very choice of possible-worlds semantics for modalities. This is because for any individual, there is *some* property (no matter how inaccessible, complex, wildly disjunctive, etc.) that that individual has at every world--e.g., simply take the disjunction of all the conjunctions of properties respectively characterizing the individual at each world in turn.[1]

Option (i) I take to be a nonstarter, for familiar Kripkean reasons. There simply are no *accessible* properties of mine that I could not have lacked and still been me. Option (ii) I leave officially open; *perhaps* I have an inaccessible essence, but the deliverance of modal intuition (mine, anyway) is hardly secure enough to convince me of that. I also remain agnostic about categorial properties such as being human, and being animate, though none of these individuate anyway.[2] But *at least* I have a haecceity, expressed by the open sentence "*x* = WGL," and I am certainly Lycan, = WGL, = me, throughout all possible worlds (which is not, of course, to say that I am *named* "Lycan" or *abbreviated* "WGL" there).[3]

At least two further arguments come to mind. The first is based on Kripke's and Rescher's objections to the "telescope" view of what it is to describe a possible world, endorsed in Chapter 4. We do not look at other worlds as through a telescope and discover their properties by surveying and scrutinizing them closely; rather, we stipulate what worlds we are considering. And in framing a counterfactual hypothesis for such a stipulation, I can use an unadorned proper name, such as "Nixon," and suppose that its bearer, Nixon himself, has such-and-such a property at the world being described, without thereby committing myself to any further nontautologous accessible characteristic of Nixon at that world. Thus for any nontautologous accessible property P, Nixon might lack P at that world, and so P is not essential to Nixon; he has no accessible qualitative essence.

The second argument is Chisholm's (1967) well-known Adam/Noah thought experiment, in which we start with an Old Testament world and gradually switch the respective properties of Adam and Noah, until Adam has all of Noah's properties except, naturally, that of being Noah, and vice versa. It seems clear enough that Adam could have been born rather than created whole, and years later; he could have had Ham, Shem and Japhet as sons rather than Abel and you-know-who; and so on indefinitely.[4] There may be a limit to this process--e.g., if Adam could not have had his genetic code completely rearranged and still have been Adam--but if so it is not an "accessible" limit in the sense introduced above.

Concession: From the fact that Adam might have lacked any given accessible property and still have been Adam, it does not *follow* that he could have lacked all or even lots of them. And indeed Searle (1958), Wiggins (1967) and D.M. Armstrong (in conversation) have all insisted strongly that even if any given property may go, an individual must preserve at least a vague preponderance of its accessible properties overall; at some imprecise but objective point in the switching process, Adam would cease to be Adam and become someone else, either Noah or someone intermediate as between the two. For similar reasons, Searle, Wiggins and Armstrong might resist my first argument from the falsity of the "telescope" view.

I have four objections to the Searle-Wiggins-Armstrong position.

(a) I think Kripke's (1972/1980) parallel arguments against Searle's "cluster" theory work almost as well as do the original arguments against Russell's Description theory. Someone who wants to maintain "cluster" essentialism will have to answer those arguments.[5]

(b) Though "Adam could have lacked P_1" and "Adam could have lacked P_2" do not jointly entail "Adam could have lacked P_1 and P_2," I think they create a presumption; they warrant believing that Adam could have lacked P_1 and P_2, unless there is some conceptual or causal or other relevant connection between P_1 and P_2 that explains why Adam could not have lacked both. (E.g., no one could keep a determinable but lack every one of its determinates.) Now, why could Adam not have lacked all his accessible properties? I suspect that what bothers Searle, Wiggins and Armstrong is precisely (what is true) that this would make him completely inaccessible, i.e., unrecognizable as Adam. There would be not the slightest indication of his real identity. In-principle unrecognizability brings out the verificationist in anyone. And the completeness of the Adam/Noah switch also stirs fears of spookstuff, viz., of the migrating ectoplasm or invisible ego or whatever it is that used to reside with the "Adam" figure but now resides with the "Noah" figure.[6] But verificationism is false; and there is no ectoplasm in Chisholm's story--a haecceity is only a property, not a stuff or a thing.[7]

(c) Let us again unleash modal intuition, with the Searle-Wiggins-Armstrong "cluster" view in mind. Consider your own case. You could have been born on a different continent. You could have been born in a different month, probably a different year. You could have grown up in a different culture; you could have married someone really exotic; you could have pursued a very different occupation; etc., etc. Now, does it not seem *almost* as obvious that you could have done all those things together? Indeed, just the first one of them makes many of the others vastly more likely, the very opposite of inhibiting them.

(d) Lewis (1986, p. 231) offers a more telling variant of my "purple chimpanzee" example from section 2 of Chapter 3 (ironically, in the context of an argument against Haecceitism):

I might have been one of a pair of twins. I might have been the first-born one, or the second-born one. These two possibilities involve no qualitative difference in the way the world is.

Indeed they do not. Yet the worlds are distinct, and in virtue of the twins' identities being switched.

In w_1, Lewis is the first-born twin, but in w_2, he is the second-born twin. Now we may suppose that at each world, the twins' lives diverge drastically from each other; perhaps they are orphaned and separated soon after birth, adopted by families on different continents etc. Then Lewis in w_1 is transworld-identical[8] with the second-born twin in w_2, and Lewis in w_2 is transworld-identical with the second-born twin in w_1. Yet in neither case does Lewis share any significant cluster or "weighted most" of properties with his counterpart at the other world. (Our own actual Lewis may or may not share one with any of the other characters.) The intelligibility of this scenario makes it a firm counterexample to the Searle-Wiggins-Armstrong thesis.

For these reasons, I shall continue throughout this book to assume the truth of Haecceitism for actual individuals.[9] Perhaps I should call my position "Agnostic" Haecceitism, since I do not rule out inaccessible qualitative essences, but I will not assume that anything has such an essence. Agnostic Haecceitism, then, is the view that: No individual has an accessible qualitative essence; every individual has an haecceity; individuals may or may not have nonunique categorial or sortal properties essentially; and individuals may or may not have unique inaccessible qualitative essences.[10]

2. A VISIT WITH QUINE

Quine wrote "Reference and Modality" (1953/1963) four decades ago, against the backdrop of Logical Positivism. The Positivists had demystified alethic necessity by identifying it with verbal analyticity, which was said in turn to be truth-by-convention; they had no truck with possible-worlds semantics; and, well in the grip of Russell's theory of reference, they had never heard of rigid designation. Let us therefore revisit Quine's paper and see how its argument fares in light of more recent Kripkean developments.

Notoriously, Quine had already attacked even analyticity itself as a specious notion (details in Chapters 11 and 12 below), but for purposes of his assault on quantified modal logic in particular, he now waived the analyticity issue and *granted* the Positivist notion of necessity-as-analyticity for the sake of discussion, noting only that on that view "necessary" is a predicate of whole sentences and applies to them in virtue of linguistic convention. Still he was concerned to argue that the *de re* necessity ostensibly expressed by quantification into modal contexts "makes no sense" (p. 149) unless one buys into the doctrine he calls "Aristotelian essentialism" and finds deeply suspect; "so much the worse for quantified modal logic" (p. 156).

If necessity is a feature generated by language rather than of things in themselves, then it is entirely reasonable to see the "essentialness" of one of a thing's properties as simply relative to a way of designating that thing. "All cyclists are two-legged" and "All mathematicians are rational" are both analytic, let us suppose; but Lavinia, a human being who is both a cyclist and a mathematician, *herself* does not have two-leggedness essentially but rationality only accidentally, or vice versa. She has two-leggedness essentially "*qua* cyclist," if you like, and rationality essentially "*qua* mathematician," but that representation-relative "essentialism" is all the essentialism there is. The number 9 is necessarily (because analytically) greater than 7 when referred to as "9" or as "$(\iota x)(x = \sqrt{x} + \sqrt{x} + \sqrt{x} \neq \sqrt{x})$," but not when it is referred to as "The number of the planets," and there's an end on't. The necessity in no way inheres in the object, the number 9, itself. "Being necessarily or possibly thus and so is in general not a trait of the object concerned, but depends on the manner of referring to the object" (p. 148).[11]

And that is why "quantification" across the necessity operator into the numeral position, as in "$(\exists x) (x > 7)$," makes no sense to Quine; for as Linsky (1969, p. 96) puts it,

...quantification, ordinarily understood, *abstracts* from the mode in which objects are designated. '$(\exists x)F(x)$' is true or false according to whether or not at least one object satisfies the open sentence following the quantifier; but whether or not an object satisfies an open sentence is quite independent of how we refer to it, or even whether we have the means of referring to it at all.

We might think of saving the intelligibility of quantified modal logic by stipulating that an existentially quantified modal sentence is true iff there is *at least one* singular term that designates the object in question and whose concatenation with the predicate in question is necessary/analytic. (Note that *having at least one designator of that sort* is a genuine property of the object.) Thus "$(\exists x)\Box(x > 7)$" would be straightforwardly true in virtue of the (granted) analyticity of "$9 > 7$" and/or "$(\iota x)(x = \sqrt{x} + \sqrt{x} + \sqrt{x} \neq \sqrt{x}) > 7$." But that suggestion would trivialize the existentially quantified modal sentences by making them too readily true: For any object O and any property F that O has uniquely, the sentence "The F is F" is analytically true and so licenses the formula "$(\exists x)\Box(x$ is $F)$," even when the narrow-scope quantification "$\Box(\exists x)(x$ is $F)$" is false as it virtually always is.[12]

We might then think of restricting the *type* of singular term that would be needed to make a quantified modal sentence true. That is, we might stipulate that an existentially quantified modal sentence is true iff there is at least one singular term *of special kind K* that designates the object in question and whose concatenation with the predicate in question is necessary/analytic. But how to motivate such a restriction? Why should "9" and "$(\iota x)(x = \sqrt{x} + \sqrt{x} + \sqrt{x} \neq \sqrt{x})$" be counted as privileged terms of "kind K" and "The number of the planets" not qualify?

At this point Quine claims that quantifying across modal operators commits one to Aristotelian essentialism, roughly the doctrine that an individual has some of its properties necessarily or essentially but others only contingently or accidentally, "quite independently of the language in which the thing is referred to, if at all";[13] Quine regards quantified modal logic's commitment to essentialism as damning, since he follows the Positivists in thinking of nonverbal necessity *in re* as an appalling "metaphysical jungle," and of the "invidious" distinction as completely arbitrary once we depart from verbal analyticities.

Four considerations conspire to support his commitment claim:

(i) The obvious difference between the first two of those designators and the third is that the first two pick out 9 by reference to its genuinely essential properties (*being 9* and being an x such that $x = \sqrt{x} + \sqrt{x} + \sqrt{x} \neq \sqrt{x}$, respectively), while the third picks out 9 by reference to a merely accidental property. But that suggestion presupposes that 9 itself has some "genuinely essential" properties and some "merely accidental" properties--what Quine (p. 155) calls Aristotle's "invidious" distinction.

(ii) The most *straightforward* solution to Quine's original problem is just to junk the idea that necessity is only verbal and regain necessity *in re*; let us revert to taking quantified modal sentences as asserting that certain objects do, themselves, have some of their properties essentially but others only accidentally or contingently. That is Aristotelian essentialism, and it makes perfect sense of the quantified modal sentences. (Moreover and accordingly, it restores substitutivity to modal contexts by allowing genuinely referential terms to occur inside them: 9 *by whatever name* is itself necessarily greater than 7. So there is a sense in which the number of the planets is necessarily greater than 7.[14]

(iii) Possible-worlds semantics provides a perfect *model* for Aristotelian essentialism. That is, a possible-worlds semanticist can very cleanly represent the distinction between a thing's essential properties and the thing's merely contingent properties; the former are those properties which the thing has at every world at which it exists.

(iv) Possible-worlds semantics also affords a precise way of picking out the special kind K of designator which will serve to license quantifying in, and effortlessly draws the connection between that kind of designator and the notion of essence: The special class in question is that of *rigid* designators in Kripke's (1972/1980) sense, a rigid designator being a singular term that denotes the same object at any world at which that object exists. One may, then, quantify in when but only when the singular term to be replaced by a bound variable is rigid; *voilà*. And what is special about rigidity that justifies this rule is that a rigid designator picks out its bearer by reference to the bearer's essence, i.e., by whatever set of properties the bearer has at every world in which it exists. But this in turn requires that there be some such set of properties, hence (again) Aristotelian essentialism.

So it would seem Quine is right; the interpretation of quantified modal logic does require Aristotelian essentialism, at least if it is to be conducted within the possible-worlds framework.

But care is needed here, for two reasons. First, as was shown by Parsons (1967, 1969), "commitment" comes in distinct grades of severity (much as did modal "involvement" according to Quine [1966b]): (a) At worst, a quantified modal logic might contain some essential sentence as a theorem. (An "essential sentence" is a formula that affirms there to exist some objects that have a property F necessarily and also some other objects that do not have F necessarily.[15]) (b) Without having any essential sentence as a theorem, a logic might entail one when conjoined with "some obvious and uncontroversial non-modal facts" (such as the existence of more than one individual [1969, p. 78]). (c) The logic might merely allow an essential sentence to be well-formed, thus presupposing the meaningfulness of the Aristotelian distinction, without in any way endorsing that sentence as true. Parsons argues rigorously that plenty of quantified modal logics avoid commitment in senses (a) and (b), though he grants that it is the very business of a quantified modal logic to "commit" in sense (c).

We have seen that there is a fourth sense of "commit," a semantical/metaphysical sense, in which quantified modal logic does by its nature commit to essentialism: The semantical apparatus of possible worlds, with the required transworld identity (or counterpart) relations, guarantees that individuals have essences of some sort, either qualitative essences or haecceities. Every individual at least has the property of necessarily being that individual, a property not shared by any other object, and perhaps has some inaccessible qualitative essence as well.

The second reason we must read Quine's commitment claim carefully is that an important qualification is required: An object's set of essential properties may be as small as to include just the object's haecceity and tautological equivalents thereof. Though a Haecceitist does distinguish between a thing's essential properties and its accidental properties, the lopsided Haecceitist version of the distinction may be felt as less "invidious" than the accessible qualitative one implied by Russellian essentialism or even the inaccessible qualitative one recommended by Kripke. Quine does not discuss Haecceitism *per se*. His examples of putatively essential properties are mainly accessible qualitative properties. But he seems to regard the Haecceitist's distinction as being invidious in the same way, for in his quick argument on p. xx176xx of Quine (1966b) he uses "$x = x$" as expressing an objectionably essential property of x.[16] Nor would it have mollified him for someone to have pointed out in a reasonable tone that I am me, = WGL, = Lycan, etc., at least at every world obeying @'s laws of nature.[17]

To Quine's attack, Linsky (1969, pp. 99-100) replies that the argument should be turned on its head; and I concur. Possible-worlds semantics *has* made sense of the Aristotelian doctrine, at least formally speaking. It has done so without deciding the question of which of a thing's properties are essential to the thing, but that is just as it should be, since logic and semantics should beg as few substantive metaphysical questions as possible. Moreover, common sense supplies apparent counterfactual truths about the tracking of individuals through worlds, and these powerful intuitions richly inform our modal intuitions about objects and their properties. That is not to resolve all disputes between Haecceitists, Kripkeans, any Russellians there may be, and others, but it does suggest there is philosophical truth to be got at, and it does render the issues attractively clear.

But why should we side with Linsky rather than continuing with Quine to take Aristotelian essentialism as a *reductio*?[18] That depends in part on one's attitude toward possible worlds and possible-worlds semantics generally: If one believes robustly in appeal to possible worlds whether Actualist or Concretist, one will agree that the truth or falsity of this or that version of essentialism is a plain fact, determined by the worlds and their contents themselves. If like me one is an Ersatzer or other wimp but preserves the formal truth of possible-worlds statements, one still can formulate essentialist claims and questions precisely in possible-worlds talk, even if one has little hope of deciding them. But if one simply rejects the possible-worlds apparatus in every sense, believing it to be all either false or incoherent or gibberish (as Quine does), one will indeed be able to make little sense of quantified modal logic--but (n.b.) not because of *quantifying in* in particular, but only because one will be able to make little sense of modal logic at all. Thus if quantified modal logic is infirm, that is not because it is committed to Aristotelian essentialism.

NOTES

[1] I assume here that there is a property corresponding to each well-formed and nonparadoxical predicate of the language, in particular, to disjunctive predicates.

[2] Could I have been a rhesus monkey? an arachnid? a ham sandwich? a ping-pong ball? a fullerine? a spacetime point? I have no stable intuitions here.

[3] For convenience I have expressed the doctrine in terms of properties, "a haecceity" being a property. But Adams (1979) and Lewis (1986, p. 225) point out that strictly speaking Haecceitism does not require the existence of properties at all; a nominalist could hold it, as a supervenience thesis dealing with, say, sentences' truth-values.

[4] Actually the thought experiment goes back at least to Wilson (1959), though Wilson finds the "biography-swap" idea absurd, and so is no Haecceitist. In commenting on Wilson, Prior (1959-60) called attention to a class of properties besides haecceities that would *not* be swapped in a case of this kind: complex ones involving the haecceities, e.g., "building an ark and not building an ark without being Noah."

[5] Searle (1983, Ch. 9) discusses Kripke (1972/1980) at some length and defends his own "descriptivism" against some of Kripke's objections, but he confines his critique to Kripke's Causal-Historical account of referring; he does not either take on the Millian semantic thesis that names are merely tags with no descriptive meanings, or defend "accessible" qualitative essentialism. (The mutual independence of the Causal-Historical theory and the Millian semantic claim is well emphasized by Devitt [1989].)

[6] The latter concern was expressed to me by Pavel Tichy.

[7] Forbes (1985, Ch. 7) and Lewis (1986, pp. 245-46) both resist Chisholm's example, in the context of counterpart theory--Forbes by making counterparthood a matter of degree and assimilating Chisholm's argument to the Sorites, Lewis by making the counterpart relation intransitive.

[8] Or counterpart; the present point does not turn on the issue of identity *vs.* mere counterparts.

[9] Lewis (1986, sec. 4.4) mounts an intricate attack on Haecceitism. But his argument simply presupposes Concretism--quite fairly, from his point of view, since he had already argued the superiority of Concretism to Actualism in his

previous chapter--so I need not delve into the details. Small wonder that Lewis opposes Haecceitism, for we saw in Chapters 2 and 4 that from the Concretist point of view haecceities are barely even intelligible, and Lewis agrees: To express Haecceitism, a Concretist would need "a *non-qualitative counterpart relation*--that label will do to be going on with, though really I regard it as a contradiction in terms" (p. 229).

My own pro-Haecceitist arguments in this chapter are intended neutrally as between Actualist and Concretist foundations; so, if they succeed on their own terms, I shall have Tollensed Lewis' Ponens, and scored a heavy blow against Mature Lewis.

[10] In one way, it is more daring and ambitious to accept inaccessible qualitative essences than to reject them, viz., it commits one to a stronger form of essentialism; granted that every coherent open sentence expresses a property, the owning of haecceities is virtually a truth of logic. But in another way, the rejection of (even) inaccessible qualitative essences is more daring, for it countenances more, and more exotic, possibilities.

[11] In fact, Quine argues that for this reason the Positivists' analyticity approach to necessity *conflicts* with the detested Aristotelian essentialism (p. 155): for the Aristotelian essentialist, being necessarily or possibly thus and so *is* a trait of the object concerned. This is one of Quine's several nice *ad hominem*s against Carnap.

[12] I would think that this same objection would succeed against Marcus' (1962) proposal that we interpret modal quantification substitutionally; at least, I do not see offhand why it does not. Incidentally, I am puzzled by a remark of Linsky's (1969, p. 97): "Presumably, Mrs. Marcus interprets universal quantifications as asserting the truth of all substitution instances of the relevant matrices." Surely not, since on that policy no universal modal quantification would ever be true; every object has *some* designator whose concatenation with the relevant predicate would be non-analytic. Presumably Marcus would interpret a universal modal quantification rather as being true iff *every object* (in the universe of discourse) has at least one designator whose concatenation with the relevant predicate would be analytic.

[13] Quine (1966b, pp. 173-74). Formally, he represents the doctrine as $(\exists x)(\text{nec } Fx.Gx. \sim\text{nec } Gx)$.

[14] As Linsky says (1969, pp. 94-95), that sense is neatly predicted by Smullyan's Russellian analysis; but Smullyan leaves it completely unexplicated and so misses Quine's point quite badly.

¹⁵ This gloss hides some points of detail that Parsons shows are needed to keep "Aristotelian essentialism" from being trivialized (see Parsons [1969, pp. 75-77], and also Fine [1978] and Forbes [1986]).

¹⁶ However, Quine (1986) accepts Kaplan's (1986) distinction between a "Benign" essential property, that is possessed by every individual for purely logical reasons, and an "Invidious" essential property, possessed by one individual but not by another; presumably $\lambda x(x = x)$ is counted as Benign rather than Invidious.

McMichael (1986) draws further distinctions within these categories. An "interesting" essential property (p. 33) is one "that individuals have...in virtue of their own peculiar natures...[rather than] merely in virtue of general necessary truths"; thus it "must not be merely the result of 'plugging up' a relation, such as identity, that is necessarily totally reflexive..., [or (e.g.)] a disjunction one of whose disjuncts is a property of the sort just mentioned." But bare haecceities in our sense are properties of just that sort, so even they do not count as "interesting" in McMichael's sense. He is really out for inaccessible qualitative essences.

¹⁷ Quine (1977, p. 8) writes,

Talk of possible worlds is a graphic way of waging the essentialist philosophy, but it is only that; it is not an explication. Essence is needed to identify an object from one possible world to anther.

Here again, he does not address the Haecceitist option.

¹⁸ Barry Loewer has pushed me on this point.

CHAPTER 6

FICTION AND ESSENCE

Lycan and Shapiro (1986) constructed a system of possible worlds, out of our familiar if not unproblematic abstract entities, identifying "nonactual worlds" with certain sets of structured propositions--including singular propositions--and identifying "nonexistent" individual objects with certain complex properties. In our formal system Shapiro and I accommodated the transworld identity, in the strict sense, of actually existing individuals: Since "worlds" are just sets of propositions, it is easily determined which individuals inhabit a given nonactual world; individual i is in world w iff w has as an element some proposition singularly about i (that is, some proposition of the form $<i, \lambda x(Fx)>$.) Thus, actually existing individuals are transworld-identified directly with themselves, not with sets of properties *or* with mere counterparts; we embraced Haecceitism.

1. THE CONSERVATIVE LINE

We wanted to treat the transworld identity of "*nonactual* individuals" differently, since neither Shapiro nor I would countenance their existence in any Concretist sense. To say that "the present king of France could exist" is to say only that it is possible for there to be a present king of France, which for us is in turn to say that some possible world has as a member the proposition "$(\exists x)(x$ is now king of France)." Unlike actuals, we said, merely possible individuals do have qualitative essences--individuating properties that necessarily apply to at most one individual per world apiece--and "accessible" ones at that.

Let us call these qualitative essences *characteristic properties*, or "characteristics" for short.[1] In the Lycan-Shapiro system, in fact, nonexistent possibles are identified with their characteristics. Moreover, the characteristics at each possible world are stipulated. A given individual's characteristic serves to pick out that individual at other worlds. Suppose, for example, that P is a characteristic at world w, and thus that "$(\exists x)Px$" is a member of w. Suppose also that another world w' contains "$(\exists x)(Px$ & $P'x)$" for some property P'; then we would say that the merely possible individual identified by P in w has P' in w'.[2] Thus Shapiro and I pejoratively distinguished the individuation of nonexistents from that of actuals.[3] We embraced traditional essentialism regarding the former despite our Haecceitism regarding the latter.

I now doubt the wisdom of that policy, having since found what I think is powerful motivation for Haecceitism regarding nonactuals. I am not entirely convinced either way, so I shall continue to take the Lycan-Shapiro line seriously; call it the Conservative Position. In this chapter I shall expound a defense of the Conservative Position, but go on to argue the opposing view as well and show how that view resists the Conservative arguments. Here, then, comes the Conservative case.

2. OPENING DEFENSE OF THE CONSERVATIVE POSITION

Contra Concretism, it would seem that nonactual individuals differ sharply from actual ones: (i) Actual individuals are given; by definition, they are part of the world we live in. They can be perceived and ostended. (And I assume that the proper names of actual individuals neither abbreviate flaccid descriptions nor are equivalent to them in any other semantical way; see Chapter 7.) By contrast, mere possibilia are not given or (*pace* Routley [1980]) encountered, but specified, stipulated, or constructed. Moreover, (ii) the only way to specify a nonexistent is to proffer a description--"the golden mountain," "the present king of France," "the winged horse ridden by Bellerophon," "Jimmy Carter's older brother."[4] (It is these specifying descriptions that Shapiro and I took as the characteristics.) And consequently (iii) it is arguable that nonexistents figure in no singular propositions, and that they are the objects of neither *de re* modalities nor *de re* propositional attitudes.[5] Offhand one sees no clear sense in which one can have a particular nonactual object in mind as opposed to one which is qualitatively just like it (though later on I shall take a giant step toward providing such a sense).

I pause to rebut an objection to the Conservative Position, put to me on separate occasions by Pavel Tichy and Penelope Mackie. It goes roughly as follows: Assume that (the real) Everest could have been uniquely made of gold. Then there is a world w_1 at which Everest = the golden mountain. Everest at w_1 is of course transworld-identical with Everest here at @, and on the Conservative doctrine, Everest at w_1 is also transworld-identical with the golden mountain at a third world w_2 even though the golden mountain at w_2 is not Everest there and (let us say) Everest does not even exist at w_2. Contradiction: By transitivity of identity, Everest here = the golden mountain at w_2, but by hypothesis that is false.

What is wrong with the argument is that transworld identity is not the same notion as contingent identity within a world, and so the appeal to transitivity is an equivocation. There are at least three ways of seeing the point.

(i) Transworld identity is the relation *being a value of the same world-line as*. That is, you at @ are transworld-identical with Jones at w just in case you and Jones are manifestations of the same world-line at your respective worlds. Now actually that should be, manifestations of *a* same or common world-line, because every individual at every world is a value of more than one world-line (different kinds of world-lines corresponding to different sortal kinds of item, *á la* Geach [1967] and Hintikka [1972]). A human-being-shaped world-line is a different sort of world-line from a role-shaped or officey world-line. Regarding Everest being the golden mountain at w_1, that single object is the value at w_1 of each of two distinct world-lines, a mountain-shaped one and a Russellian role- or niche-shaped one. So Everest at w_1 does share a world-line with Everest at @, and Everest at w_1 does share a world-line with the golden mountain at w_2; but it does not follow that Everest at @ shares a world-line with the golden mountain at w_2, for the foregoing two shared world-lines are distinct world-lines. Hence it does not follow that Everest at @ is transworld-identical with the golden mountain at w_2. (Doubtless on *some* crazy scheme there is a truly weird world line that has as values both Everest at @ and the golden mountain at w_2, but that is irrelevant here.)

(ii) In other perfectly ordinary cases of intra-world contingent or role-occupant "identity" the inference obviously fails. At @ David Lewis is a philosopher and George Bush is President of the USA. Let us say that at w_1 David Lewis is President of the USA, and at w_2 Irving Smedley is President of the USA, while David Lewis does not exist at w_2. The presidency is the *same office* from world to world, and Lewis and Smedley are the *same official* from w_1 to w_2, but as before no one would think of inferring that the real David Lewis is President at w_2 just because he is President at w_1. Qualitative essences for nonactuals (or for anyone) make those "individuals" into roles or offices, occupied at a world by this or that individual in the more "genuine" sense of the term "individual."

(iii) Whether one is a Lewisian Concretist-*cum*-counterpart-theorist or an Ersatzing wimp, the relation of transworld identity and intra-world identity are not the same relation. (a) The latter, but not the former, can be contingent. (b) If one is a Lewisian, there is no genuine transworld identity but only counterpartship, while there is genuine intra-world identity. If one is an Ersatzer, transworld identity is literal numerical identity for actual individuals, but "identity" within a nonactual world consists just in an identity *proposition*'s being an element of that world. (When the identity proposition is a contingent one, it is not even genuinely an identity proposition, being equivalent in Russell's way to a subject-predicate form.)

Despite my initial remarks in its defense, the tendentious Conservative Position is entirely questionable. One may be particularly dissatisfied with its assumptions that there are no singular propositions involving nonactuals and that (accordingly or vice versa) there are no attitudes or modalities *de* nonactuals. Among others, Lewis (1983) and McMichael (1983a, 1983b) have maintained that an adequate modal semantics must accommodate all these features, contrary to our assumptions;[6] let us consider McMichael's argument in particular.

3. MCMICHAEL'S ISSUE

McMichael raises a question regarding modal properties of nonactuals. It takes several different forms, but I think the most trenchant version of the objection is a simple modal one: Given such-and-such a nonexistent object introduced and universally accepted as being "the F," *could that object have failed to be F?*

Intuitions differ from case to case, and from person to person even within cases. I distinguish three main types of case: (a) "nonexistents" generated solely by definite descriptions made up on the spot, such as "the golden mountain" or "the town just north of Mayberry in which Marie Curie discovered the secret of eternal life"; (b) fictional objects already well entrenched in literature and/or in culture;[7] and (c) nonactual items endowed with their own modal properties by initial stipulation. Let us consider these *seriatim*.

Could *the golden mountain* have been made of aluminum, or of marshmallow? Could *the town just north of Mayberry...* never have been remarkable in any way at all, or have been located near Kinston? I do not know what to make of these questions. They seem senseless, or at least to have no answers, since the descriptions they involve have been introduced brutely, without context, and there is no background story or other information against which to measure modal stretch. For nonexistents of type (a), descriptive or qualitative characteristics of the sort we have posited seem entirely appropriate. However, types (b) and (c) are less straightforward. Let us turn to the case of fictional individuals; but first a small digression is required.

For clarity,[8] we must distinguish fictional *characters* properly so-called from fictional *persons* or other individuals. A fictional character, in one common use of the phrase (cf. van Inwagen [1977]), is an actual though abstract literary entity, ontologically on a par with a *novel* or a *story*. In one sense characters are proper consistuents of novels and stories,[9] and are open to description and evaluation of the same sorts as are their containing works themselves: A character may be well-drawn or a two-dimensional cardboard pinup, vital or inessential to the plot, more ingeniously or less creatively conceived than Dorothy Sayers' Mervyn Bunter, a faithful realization of the author's original intention or an unexpectedly evolved departure, popular or unpopular with readers, etc. Such things cannot (except metaphorically) be said of real flesh-and-blood people; no more can they be said of fictional flesh-and-blood people in the sense I intend. A fictional person or other individual is a nonactual possibilium having, at its own home world or worlds, the properties described in some (presumptively but not always actual) fiction.[10] Thus, fictional people are fat or thin, smart or stupid, candid or devious, and the like.

Once the foregoing distinction is enforced, the question of whether there are fictional *individuals* at all becomes controversial. Some writers, such as van Inwagen (1977), decline to treat "fictional characters" as nonexistent possibilia or as denizens of possible worlds even if they grant that in some sense there are other kinds of nonexistent possibilia; other writers (including Kripke [1972/1980, 1972b], Kaplan [1973] and Fine [1984]) deny *possibility* to fictional individuals *tout court*. I shall not join the labyrinthine issues involved here;[11] for purposes of this chapter I shall simply continue to assume that in whatever sense there are nonactual individuals, there are in particular fictional individuals, falling in with the view that actual fictions describe nonactual possibilia. If positive argument is needed, I offer just three brief points: (a) Fictional individuals such as Pegasus, Hamlet, Sherlock Holmes *et al.* have always been used as paradigmatic examples by partisans of nonexistent possibles (though I suppose such philosophers *may* have been hopelessly confused from the beginning). (b) Novels and stories generally seem to say things that could have been true even though they are not. If so, and if possible-worlds semantics constitutes our standard device for handling modal statements, then sentences occurring in fictions are true at worlds other than our own and so it seems that fictional individuals exist at those worlds. (c) As we shall see, fictional individuals appear to have modal properties, and--again according to our general practice of explicating modality in terms of worlds--this seems to show that fictional

FICTION AND ESSENCE

individuals somehow inhabit worlds. Let us now end our digression and return to this last issue.

4. THE MODAL PROPERTIES OF FICTIONAL INDIVIDUALS

Consider Perry White, fictional editor of the *Daily Planet*. Might he have gone into bricklaying, philosophy or nursing instead of journalism?[12] I am pulled two ways here. On the one hand, "Perry White" is only a fictional *character*, a creation of a human author; all there is to him, so to speak, is his two-dimensional role in the story as Clark Kent's unsympathetic, cigar-chomping boss. Had the author left out this "boss" character entirely and at the same time introduced a janitor, or an itinerant drunk who wanders in off the street, and happened to name it "Perry White," this new character would not be Perry White in the sense in which we now use that name; it would simply be a different character despite the accidental sameness of moniker.

On the other hand, as is well known, fictional characters do not have only the properties explicitly ascribed to them in their native works, but also those which may fairly be extrapolated or assumed on the basis of the text and its setting. Now, the original character Perry White is (in the story) a man. Presumably he has toes, even though we never see them and even though there is no specific toe length that he has. Similarly, like any man he presumably has a genetic code, even though there is no specific genetic code that he has. And why not suppose that he has the everyday sorts of modal properties that ordinary men have as well? Every actual man is such that he might have pursued a different occupation, so why not Perry White? In one episode or another White might even be moved to remind Clark Kent of this possibility, saying something like, "I should have listened to my mother when she told me to stay out of journalism because of idiot milquetoast jerk reporters like you!"

But if we grant (for this or any other reason) that White might have joined a different trade, it seems we are saying that at a world other than that described in the Superman stories, White does go into bricklaying or whatever instead of editing the *Planet*. And this conflicts with our assumption about his qualitative essence. If his editorship is not a reliable transworld identifying mark, the same can be said of any other feature attributed to him by his creator; and so it seems we must award him a haecceity, just as if he were actual. Yet this would be extremely problematic, as our original argument for qualitative essences implies and as I shall further consider below. (Can we imagine a story just like the *Superman* corpus except that in it Perry White's and Clark Kent's haecceities are switched??)

It might help to distinguish two sorts of truth, in a thoroughly familiar way: truth *simpliciter* from truth-in-fiction. I want to say it is not *true* (period), even though it is indisputably true-in-fiction, that there is a person (Clark Kent) who can leap tall buildings at a single bound.[13] If this distinction is sound, it facilitates the accommodation of our occasional Haecceitist feelings about fictional characters. It is indisputably true-in-the-comics that Perry White has toes, genes and modal properties. It is not true in reality--or, less tendentiously, true *simpliciter*--that "he" does. Whatever account we might be moved to give of truth-in-fiction, what concerns us in fashioning semantics for natural language is truth, period, and it is not true in this sense that there is someone called "Perry White" who is a newspaper editor but might have been a bricklayer. What is true--at the outside--is that there is a character, Perry White, who is (designated as) Clark Kent's editor, and that is all there is to it; he has modal properties only in-the-story.

Unfortunately the matter cannot be settled so easily. For I have granted at least that White has his modal properties *in-fiction*, and as in Chapter 3, the semantics of this "in-fiction" operator remains to be determined. When provided, it may be found to cause trouble for my brand of actualism. Indeed we may expect it will, for the only natural possible-worlds interpretation of "It is true-in-fiction that P" is something like, "It is true that P at every world compatible with the implications and presumptions of the (relevant) story,"[14] and *this* seems to establish one or more worlds at which White becomes a bricklayer.

5. MCMICHAEL'S ARGUMENT

The objection is reinforced by an argument of McMichael's that is not tied to the vexing vagaries of fiction but is based on plain iterated modalities of type (c). Again (almost) indisputably, there might have been someone having such-and-such a property F that also had but might have lacked a further property G; for example, Rose Kennedy might have had a son, distinct from Joe Jr., Jack, Robert and Ted, who went into philosophy but who might instead have gone into bricklaying. Thus on anyone's possible-worlds semantics, fiction and its awfulness completely aside, it seems there is a world w containing the supernumerary philosopher Kennedy and a further world w' also containing that very Kennedy but at which he went into bricklaying instead of philosophy--not because of any work of fiction but simply because the envisioned possibilities seem genuine. Thus we are forced to consider transworld identity-conditions for McMichael-individuals.

We might naturally try to assimilate such individuals to cases of type (a). That is, we might insist that the imagined philosopher son of Rose Kennedy must remain a philosopher throughout all the worlds he inhabits just as the golden mountain must remain golden, since he has in effect been stipulated on the spot in the same way. But this would be to deny the truth of our original iterated modal sentence, and that sentence still seems true (in reality, not in fiction of any sort). We must grant that "Rose Kennedy's fifth son" is a philosopher at some worlds and not at others. We can get away with granting that, so long as we are allowed to split the original description "Rose Kennedy's fifth son who is a philosopher" into essence and accident: let that individual's essence be simply "x is Rose Kennedy's fifth son," abbreviated "$R(x)$." One world ($\neq @$) contains $R(x)$ as an characteristic and "$(\exists x)(R(x)$ & x is a philosopher)" as an element; yet another world containing $R(x)$ has "$(\exists x)(R(x)$ & x is a comedian)" as an element instead.

But if the philosopher son is only contingently a philosopher, might he not also be only contingently a *son*? It would seem that just as there might have been an F (distinct from every actual individual) who is but need not have been G, there might have been an F (ditto) who need not have been F. Consider "There might have been someone in the doorway who need not have been in the doorway." If context provides an extrinsic essence for the supposed person (say, "the person I am about to say hello to"), there is no difficulty, but if no such extrinsic essence is ready to hand, the person is stuck in the doorway at least *pro tem*. If we are to respect these iterated singular modalities in the absence of "extrinsic" essences, it seems we must move to nonactual haecceities of some sort after all; we must grant that there is a property of *being N*, where N is a proper name of our imaginary philosopher, that persists from world to world despite variation of all his ordinary features.[15] If there is a nonqualitative property of *being Perry White*, for example, and a nonqualitative property of being Clark Kent, then there is after all a world just like that of the Superman stories in which everything is the same except that White has all of Kent's qualitative features and vice versa.

Numerous objections may be brought, and have been brought, against this idea of nonqualitative haecceities for nonexistents. The Conservative Position still has plenty of fight in it.[16] Yet I think that the objections can be circumvented, albeit with some considerable effort, and that there is more to be said in support of such haecceities than might at first appear; so in what follows I shall offer, to those theorists who do insist on taking the iterated modal properties seriously, a Haecceitist alternative, since I maintain that the idea is coherent and defensible even if it seems unattractive.[17]

6. MORE ON BEHALF OF THE CONSERVATIVE POSITION

Let us state more fully the main objections to Haecceitism for nonactuals. One is that with which I began: that nonexistents can be introduced only by description.[18] A second (closely related) is that even if we can coherently think of a world just like this one save for the switching of Adam and Noah, we simply can*not* distinguish two worlds which differ only in the switching of the alleged haecceities of Rose Kennedy's philosopher son who might have gone into bricklaying and her seventh, bricklayer son who might instead have gone into philosophy, or (cf. Adams [1981]) two worlds which differ only in the switching of two nonactual electrons.[19] Third, if nonactuals have haecceities and can differ numerically without differing qualitatively, then it ought to be possible for us to have a particular object in mind, or to have a propositional attitude toward that object, without having in mind a qualitative twin; yet this does not seem possible.[20]

Fourth, as Adams observes (1981, pp. 11-12), an ordinary haecceity (assuming real people have them) bears a special relation to its owner: it could not exist without its owner's (that very person's) existing also. But if actualism is correct no nonactual "individual" can enter primitively into any relation, and in particular there can be no state of affairs which consists simply in Perry White's exemplifying Perry-Whiteity. To put the point slightly differently: Predicates which express the haecceities of actual individuals are syntactically and semantically complex, consisting of the identity sign concatenated with a rigid designator whose reference has been fixed by a process involving causal contact with the referent itself; but a "nonactual individual" is not genuinely an individual on our view, and cannot stand as a term in a genuine identity predication. Fifth, it is not easy to see the difference between an "unexemplified haecceity"[21] and a good old Meinongian *nonexistent possible*, once there is no visible residue of familiar qualitative properties to go proxy for the noxious Meinongian object; McMichael (1983a, p. 61) asserts that the introduction of haecceities for nonactuals "seems tantamount to acceptance of possibilism." Sixth,[22] consider a single possible world containing two planets qualitatively identical to each other--say, each planet contains a replica of Conan Doyle's Victorian England and in particular a Sherlock Holmes figure. What conceivable ground could there be for choosing one of the two Holmes figures and deciding that it is the real Holmes while the other is only its qualitative duplicate? To attach a Holmesian haecceity to either figure would be entirely arbitrary.

All six of these objections have force;[23] that is why Shapiro and I fashioned our official semantics without recourse to unexemplified haecceities.[24] But for those who want to take the putative modal properties of nonexistents more seriously, I now offer an approach that will take at least some of the sting out of our objections to haecceities for nonactuals.

7. YES, HAECCEITIES FOR NONACTUALS

Most of the objections stem from the fact that nonactual individuals stand in no causal relation to us and are known only by description. But it is possible to frame a Causal-Historical theory of reference, even for empty singular terms, that affords a finer-grained individuation scheme. Consider fictional individuals again. At least part of what makes Perry White the person he is are the circumstances of his character's creation. We might say that a fictional person qualifies as being (identical with) Perry White if and only if the relevant use of that person's name is connected in the right historical way with Jerry Siegel's original act of writing (in the real world).[25] Several sorts of cases fall nicely into line with this suggestion. (i) If some writer or cartoonist in Apex, North Carolina, who has miraculously never heard of the Superman comics, just as miraculously happens to invent a character also named "Perry White," that character is not the same character as Mr. Siegel's Perry White, even if he or she is similar in remarkable respects to Clark Kent's editor.[26] However, (ii) if someone undertakes to write a spinoff strip about the original Perry White's journalistic triumphs and tribulations or about his private life, with the original firmly in mind and with the intention of grafting the new stories smoothly onto the old--as Ruth Plumly Thompson continued L. Frank Baum's Oz books with Baum's explicit imprimatur, and as George Macdonald Fraser has brilliantly written about Thomas Hughes' Flashman--I am willing to count the new White as being the same character as the old.[27] (iii) Mr. Siegel could have written a new Superman story in which Superman (or another character) discovered that the *Planet* editor was an imposter, and that the real Perry White had years ago decided against journalism and gone into philosophy instead (but changed his name for reasons of euphony to "Hilary Putnam"). This would stick, it seems to me; after the new story had been published it would be true-in-fiction that Perry White was really a philosopher and had only erroneously been thought to be Clark Kent's editor.[28]

If some such Causal-Historical criterion could be made to work, it would provide a means of distinguishing qualitatively identical characters and so lessen the pressure of our objections to unexemplified haecceities. The initial problem was that nonactuals are identified only by description because they are causally unconnected to us; but as we have seen, their names, their dubbings, and subsequent acts of referring or "referring" to them are not so unconnected. Potentially they can be used to distinguish descriptively identical characters. If this conjecture is sound, our first objection fails.[29] So does our second; for transworld identities in the "nonactual" analogue of Chisholm's "Adam"/"Noah" example can be established similarly, when pegged to real-world fictive acts and authors' intentions. In particular, we can extrapolate Kripke's emphasis on *stipulation* to the case of nonactuals. What distinguishes the White-and-Kent-switched world from its progenitor is simply stipulation backed by the right historical connection and the right intentions, just as for existents.[30] The same can be said even for McMichael-individuals: Rose Kennedy's philosopher son is himself at another world, even if he has shed his philosophizing for bricklaying and so on, if the utterer of his original supposition so stipulates under the right conditions. Our third objection goes wide as well. Odd as it may seem, I can want Pegasus rather than a descriptively identical winged horse, if I know that the latter nonexistent's name (homonymously "Pegasus") has a different ancestry from that which figures in the classical myth, e.g., if it was made up last week by my highly and coincidentally imaginative plumber.[31]

Our fourth, fifth and sixth objections are harder to turn aside, even with the aid of our Causal-Historical idea, but we can say at least a bit in response to each of them, in turn: Regarding the fourth, Adams is bothered by the fact that an ordinary (exemplified) haecceity is identified by reference to its owner and could not exist in the absence of singular states of affairs involving that owner, while no actualist ontology can provide a genuinely singular state of affairs involving a nonexistent. But the Causal-Historical account can help at least with the identification problem; the haecceity of a nonexistent could now be identified by reference to a fictive act or other quasi-dubbing. Moreover, we do not have to suppose that a haecceity exists in the absence of its owner, for we can say that at the relevant nonactual world, its owner does exist--the presence of the haecceity and the existence of its "owner" are one and the same state of affairs there. (Neither the robust Haecceitist nor the Lewisian Concretist will be satisfied by this account, of course; the unexemplified haecceities and "nonexistent individuals" it provides are pale shadows or feeble imitations of real Haecceities and individuals. That is as it should be, for an actualist. All I claim is that the account is itself formally coherent and affords a tenable identifying criterion for unexemplified haecceities.)

Regarding the fifth objection, the question of what advantage unexemplified haecceities have over bare Meinongian possibilia, I would point to the one key difference: I have argued that the real trouble with Meinongian possibilia is that (a) they force us to disambiguate the existential quantifier as between actual existents and nonexistents, but that (b) at the same time we are allowed no expressible means of doing so; the Meinongian merely asserts that the quantifier *is* ambiguous and arrogantly declines to discuss the matter further. By contrast, an unexemplified haecceity is an actual item, a property, which does not differ in ontological status from any other property despite being "blank"--it is part of the transmundane framework. (What marks it as exemplified at a world where the "corresponding individual" exists is just the relevant existential proposition's being an element of that world; thus I sidestep Adams' objection. Of course, these "haecceities" for nonactuals are not like those of real individuals, since they are not composed of the identity relation concatenated with the individuals themselves; they are watered down.[32] That is to be expected from Ersatzers.)

Regarding the sixth and final objection: Here again we rely on Kripke's seemingly justified penchant for stipulation of the identities of individuals at other worlds. Which "Holmes" really is Holmes depends on our stipulation in the real world, provided that our stipulative act is connected in the right historical way with Conan Doyle's writing of the Holmes stories.[33] If our stipulative act is *not* connected in that way with Conan Doyle's fictive act, then we cannot attach Holmes' haecceity to either of the "Holmes"'s, and we are not describing a world containing the authentic Sherlock Holmes at all. (Incidentally, the double-"Holmes"ian world is actually an embarrassment to qualitative essentialism regarding fictional characters, more so than to an Haecceitist view. For essentialism implies that qualitatively identical individuals are one and the same individual. Thus all there would be to distinguish two "Holmes"'s in the double-Doylean world would be their relations to each other's planets, provided that the planets were themselves distinguished by containing actual individuals as well as the nonexistents proper to them.[34] This may be coherent, but is not very attractive; and there is no way that I can see to decide which "Holmes" is Holmes--yet on the qualitative-essentialist view, at least one of them would have to count as Holmes, unless Conan Doyle had resourcefully avoided future counterexample by stating explicitly that there was no other planet containing a Holmes-replica.)

Thus a Haecceitism for nonactuals can be sustained, if not loved.

8. TWO FURTHER OPTIONS

There are two further possible ways of accommodating modal properties of nonactuals. Both involve denying that nonactuals can be literally identical across worlds. One is to make all nonactuals worldbound, and import a counterpart relation to replace transworld "identification." This is of course Lewis' (1983) choice, and in effect Morton's (1973) and McMichael's (1983b);[35] each author suggests some sparse structural axioms governing the posited relation. This goes against my constructivist grain; I could not in good conscience take something so crucial as primitive. (And I have no idea how the relation might be defined without thereby giving rise to a qualitative essence; remember again my contention (fn 16) that iterated modalities pose just as nasty a problem for Lewis as they do for actualists.)

The final option is to bite a very obdurate bullet and deny flatly that there is any way *at all* to identify nonactuals across worlds[36]--our characteristics would serve to identify individuals within worlds but never across worlds. This would make our iterated modal predications *essentially false*: necessarily, Perry White could not have been a philosopher, nor Rose Kennedy's fifth son a bricklayer, nor the person in the doorway somewhere else. Period. But I see no advantage of this option over the Conservative Position.

As I said above, if we are to eschew unexemplified haecceities and unexplicated counterpart relations alike, we need to explain away the contrary Haecceitist intuition. What I think the Conservative theorist must say is that (some) people's imaginative faculties are prone to supply definiteness and determinacy where there is none. If the name "Smedley" is introduced as being that of a possible fat man in the doorway, one's imaginative faculty may ignore the fact that no determinate genetic code nor any other true individuator has been supplied along with the bare description, and suppose that an individuator has been fixed even though we do not know it. *On this false supposition*, we may then accordingly (seem to) attribute modal properties to the "individual." But since there is in fact no individuator, the modal properties are unreal.

9. GEACH'S PUZZLE ABOUT INTENTIONAL IDENTITY

It may be and should be wondered what my account has to say about Geach's (1967) problem of "intentional identity." Crudely, the problem is that a nonexistent individual may be the topic of each of two different propositional or other psychological attitudes--attitudes of the sort we would usually think of as *de re*, even though there is no actual *res* toward which the attitudes are directed. Geach's example is

(1) Hob thinks a witch has blighted Bob's mare, and Nob thinks that she (the same witch) has lamed Cob's sow.

Another example might be

(2) Gonzo fears there is a burglar at the door and hopes that that very burglar has not stepped on his turtle.

Each of these sentences is open to any number of interpretations, some of which are obviously unproblematic from anyone's point of view. But according to Geach there is a further sort of reading which resists being displayed by any standard logical means: that on which both of the attitudes in question are directed toward *nonspecific* individuals, yet in some sense their foci are *the same*. E.g., Gonzo fears that there is a burglar at the door (that is all he fears--there is neither any actual individual nor any particular pre-designated nonactual such as Raffles that he suspects of burgling); yet he hopes that *that* burglar has not stepped on his turtle, not just that there is no burglar who is at the door and has stepped on it, or that there is no burglar who he fears is at the door and has stepped on it, or the like.

Geach's problem seems to me cognate with Lewis' and McMichaels', in that it turns on the transworld identity of nonactuals who are not presumed to share their most salient properties. It is another case of a nonactual individual introduced by one modality, which is said to have further modal properties (here, nonalethic *de re* attitudinal properties) in its own right: (1) says that Hob thinks there to be a witch who did so-and-so, which witch is also thought by Nob to have yet another attribute; (2) says that Gonzo fears there to be a burglar, which burglar is also hoped by him to be a non-chelonicide. The natural possible-worlds semantics for each sentence takes us to one or more worlds in which there is an individual of such-and-such a description ($\neq$ any existing individual) and then to another world in which *that same* individual has some further property yet (perhaps) may not satisfy the previous description, thus forcing our earlier question of transworld identity for nonactuals.

If Geach's problem is cognate with Lewis' and McMichael's problem in this way, that would predict a corresponding intuitive reaction on our part. In response to Lewis and McMichael the Conservative theorist doubted that the putative iterated modalities really had sense, on the grounds that (roughly) the initial hypotheses determine only generic individuals that are not nonqualitatively distinguishable; the individuals do not, except possibly by virtue of further imaginative and stipulative activity on our part, have any qualitative or modal reach beyond what is provided by the initial hypotheses. The same is true of Geach-individuals, and so if we slam ourselves back into Conservative mode, we should expect to find ourselves with the same suspicion of senselessness.

Prediction confirmed: I do find myself with that suspicion. Once the various unproblematic readings have been set aside, I do not hear any natural or intuitive and coherent interpretation of (1) or (2). (The same is argued by Dennett [1968].) Indeed, I have more trouble hearing such an interpretation for examples of Geach's sort than I do for McMichael-individuals. And when we consider the matter in terms of possible worlds, an explanation of this difference in tractability readily presents itself: McMichael's examples begin with statements of *possibility*, which semantically are associated with existential quantifiers--at *some* world there is some individual which has such-and-such a further modal property. But Geach's begin with propositional-attitude verbs, which call for universal quantification over worlds (cf. Hintikka [1962])--at *every* world compatible with what Hob thinks, there is some individual which has the further attitudinal property. In Geach's case we are explicitly forced to envision a multiplicity of worlds, each inhabited by a witch, the various witches being quite different and presumably distinct from each other across the various worlds. This makes it even harder intuitively to attach sense to the suggestion that Nob's thought picks out any one of these different witches to the exclusion of the others, and so harder to understand how Nob could in any sense have in mind "the same" witch as Hob has. But if nonactuals have haecceities, we should expect that Geachian sentences might have genuinely *de re* interpretations, the haecceities being there all along to serve as the attitudes' *rei*.

Thus an equivocal conflict between the Conservative Position and Haecceitism over Geach's issue: If one accepts the strong readings that Geach demands for his sentences, (a) the Conservative view is embarrassed by its lack of any obvious way of providing those readings, and (b) Haecceitism is strongly suggested in the way just noted. But if for Hintikkan reasons one rejects the strong readings, then correspondingly Haecceitism is discouraged and Conservatism is confirmed. And at the same time, pre-existing allegiance to the one policy or the other powerfully affects one's pro- or antiGeach intuitions.

But with only modest ingenuity, sense can be supplied to (1) and (2) on at least one more subtle Geachian understanding than the unproblematic ones. For if Hob's witch can be pegged to his own original mental act, then Nob's witch may be pegged in turn to Hob's witch in virtue of the (presumed) historical connection between his (Nob's) mental act and Hob's. The trouble will be in getting Hob's "witch" pegged in the first place, because of our semantical mandate that Hob has a multiplicity of equally appropriate witches, but a sufficiently clever and detailed Causal-Historical criterion might be able to surmount it and deliver a Haecceitist account even here. At least, the strictly Conservative account can be avoided and Hob's witch allowed her cross-modal properties.

Yet even on a Causal-Historical view of things, full-bore Haecceitism is blunted in Geach's cases by Hintikka's point. Look just at the first conjunct: Hob thinks *some witch or other* has blighted Bob's mare. That indefinite NP does not create a haecceity for Nob's thought then to latch onto, even if it does introduce a role-shaped world-line that (1)'s pronoun latches onto, nor does anything else in the conjunct create one; there is no singular reference in the clause, not even to a nonactual. Perhaps, then, Geach's examples do not militate either way as between Conservatism and Haecceitism.

10. EDELBERG'S ASYMMETRICAL CASES

Edelberg (1986, forthcoming) has carried the "intentional identity" literature an important step further. Consider his ingenious example:

...Arsky and Barsky jointly investigate the apparent (but not actual) murder of *Smith*, and they jointly conclude that he has been murdered by a single person, though they have no suspect in mind. A few days later, they jointly investigate the apparent (but again not actual) murder of *Jones*, and together they conclude that Jones was also murdered by a single person. At this point, however, a disagreement ensues. Arsky thinks that the two murders are completely unrelated (and that the man who murdered Smith, but not the man who murdered Jones, is still in Chicago). Barsky, however, thinks that one and the same person murdered both Smith and Jones. (forthcoming, p. 26)

In this scenario, (3) seems true:

(3)   Arsky thinks someone murdered Smith, and Barsky thinks he [the same person] murdered Jones.

But (4) seems false:

(4)   Barsky thinks someone murdered Jones, and Arsky thinks he [the same person] murdered Smith.

Thus intentional identity (if we grant it is real) is asymmetric!

Naturally, the Conservative Position cannot handle this case any more than it could Geach's original one. But as before, if one has qualitative-essentialist intuitions one will insist that (3) is true only on some of its deflationary interpretations and that the intended interpretation is incoherent. And even if one does not have qualitative-essentialist intuitions, there is the problem of multiple murderers, different ones at Arsky's various doxastic alternatives. But suppose one is a Haecceitist and also finds (3) and (4) perfectly intelligible.

For Haecceitists in Geach's case, I suggested that a Causal-Historical theorist might peg Hob's witch to Hob's original mental act, and then (somehow) peg Nob's witch in turn to Hob's, even though that would not seem to create a full-fledged haecceity. But Edelberg's example is not so simple, for it adds the extra problem of asymmetry. No single haecceity would do. Even if we were able to latch Arsky's belief onto a particular haecceity H, and then latch Barsky's first belief also onto H (in order to make (3) true), Barsky's second belief (reported in (2)) would presumably also latch onto H, and (4) would come out true rather than false.

The only way to avoid this collapse would be to suppose that either Arsky or Barsky or both were doxastically related to each of *two* haecceities rather than just one. For Barsky that suggestion is implausible, though I can *imagine* a theory according to which Barsky's beliefs really were *de* distinct (nonactual) people even though with believers' customary underprivileged access his internal representations of the two people were indistinguishable. A two-haecceity hypothesis is initially more plausible for Arsky, since Arsky does think that two murderers are in the field. Barsky's nonactual suspect would then be identified with Arsky's first murderer rather than Arsky's second. But then things go wrong: If Barsky is related to just the one haecceity, then sentence (4) comes out true again.

Notice that no Edelberg asymmetry can arise when the detectives' beliefs are *de* real people. Suppose Carsky believes *de re* of Giancana that he murdered Smith and Darsky believes similarly that Giancana murdered Jones. Then sentence (2') comes out true, not false.

(2') Darsky thinks someone murdered Jones, and Carsky thinks he [the same person] murdered Smith.

When the actual objects of *de re* belief are identical, the identity is perforce symmetric. What this seems to show is that neither Edelberg's example nor probably Geach's before it is a straightforward case of *de re* belief, in part because they do not involve haecceities; the resumptive pronouns that figure in such cases should receive some less ambitious treatment.[37] But I shall have to leave the matter at that. So far as I can tell, Edelberg's example is at worst neutral on our question of Haecceitism for nonactuals, which leaves that Haecceitism unharmed.

NOTES

[1] Shapiro and I originally called these individuating properties simply "essences," but in view of our earlier distinctions that usage might cause confusion here.

[2] N.b., it does not follow that P is a *characteristic* in w'.

[3] Rescher (1975, Ch. 3) elaborates and defends a similar idea of "supernumerary" individuals.

[4] For similar and more extended defenses of an inegalitarian attitude towards the individuation of nonexistents, see Kaplan (1973, 1975), Rescher (1975, Chs. III and IV), Kripke (1972), and Plantinga (1974); Robert Howell questions this argument, albeit programmatically and somewhat obscurely, in (1979, sec. 7). Somewhat different defenses may be found in Skyrms (1981) and in Adams (1981, sec. 2).

⁵ *Pace* Howell (1979), Pollock (1980), Routley (1980), Lewis (1983), and others; see the reply to objection 5.5 in section 5 of Lycan and Shapiro (1986).

⁶ Their concern is glancingly anticipated by Morton (1973).

⁷ Actually the inclusion of fictional characters as nonexistents and as inhabitants of possible worlds is neither unproblematic nor uncontroversial; see below.

⁸ I am grateful to Dick Smyth, Kate Elgin and Ken Taylor for insistent criticism on the present point.

⁹ In another sense, obviously, the constitutents of novels and stories are sentences, words, letters or the like.

¹⁰ By "fictional" here I mean *merely* fictional; for simplicity I ignore individuals who are mentioned in fiction but are also actual, and I even more emphatically ignore the hybrid people found in *romans à clef*.

¹¹ See also Castañeda (1979), Howell (1979), Parsons (1980), Routley (1980), Fine (1982), Bertolet (1984a, 1984b), and all the multifarious works further cited in these.

¹² This is not intended as a question about what Mr. Jerry Siegel (the author of *Superman*) might have chosen to *do* with the character, as when we ask whether he might have made White a health nut. Cf. my distinction between fictional characters and fictional people.

¹³ This observation strikes me as intuitively and strictly correct, but I admit I have encountered some hardy bibliophiles who maintain that Kent truly and literally works at the *Daily Planet* and doubles as Superman and so on.

¹⁴ For key refinements, see Lewis (1978).

¹⁵ McMichael offers two more subtle arguments against pinning nonexistents of his sort to "qualitative essences": that given a suitably bifurcated and symmetric world, any alleged qualitative essence might fail to discriminate an inhabitant from its doppelgänger, and that two nonactual worlds might be connected by a chain of individually small changes that gradually switch two characters' properties (cf. Chisholm's (1967) original "Adam"/"Noah" case as discussed in the previous chapter, though McMichael's own argument relies on less aggressive assumptions). I think both arguments can be resisted, by enforcement of the alternative view of nonactual-individuation that I shall sketch below, so I shall not pursue them here.

N.b., it is also open to a granite-jawed Actualist to deny after all the truth of McMichael's iterated modal sentences on the grounds that they would make good sense only on a strong version of realism about possibilia. One might imaginatively stipulate that Rose Kennedy had a philosopher son in addition to the ones we know, and count it true-in-the-stipulated-scenario that the son might have gone into bricklaying, but refuse to give a standard possible-worlds semantics in turn for truth-in-a-scenario and refuse to grant it true *simpliciter* that the imaginary son might (literally) have gone into bricklaying.

[16] Incidentally, McMichael presents his troublesome argument as a problem for *Actualism* specifically, as if the Meinongian/Lewisian "possibilist" were not to be troubled by it. Initially we can see why: it is the Ersatzer or other Actualist like me who denies that nonactual individuals are *in propria persona* constituents of other worlds, and so has trouble accounting for the transworld identity of nonactuals. But I do not see why McMichael's problem does not upon examination infect "possibilist" semantics as well. Consider Lewis' Mature Concretism. Superficially that view offers a straightforward account of singular iterated modalities: Perry White might have gone into nursing iff he has a counterpart at some other world (still distinct from @) who does go into nursing; the counterpart is a flesh-and-blood individual similar to White, just as robust a constituent of his or her own world as White is of his, and so there is no embarrassing ontological asymmetry of the Conservative sort. But, I contend, McMichael's dilemma rears its head when we ask what counterpart or similarity relation is operative. Not an effable, qualitative relation, since to set up the dilemma at all we need the assumption that White could have lacked any of the features conventionally associated with him by the stories. But presumably not a nonqualitative, *sui generis* property either, for this would be (as Lewis himself agrees on p. 229 of [1986]) to leave a mystery as the sole ground of modalities *de* nonactuals. Thus I do not see that as regards McMichael's problem Lewis is any better off than I am.

[17] Shapiro and I (1986) provided a harmless corresponding variation on our original formalism.

I am of course neglecting an option: *inaccessible* qualitative essences for nonactuals. Someone might argue that if actual individuals have inaccessible essences, then by analogy nonactuals do as well; e.g., a nonactual human being might have its genetic code as its essence. But this would create a nasty problem for the case of fictional individuals at least (anticipated by Kripke [1972/1980, 1972b]): There are countless worlds containing "Sherlock Holmes" figures, each of whose "Holmes" is realized by a genetically different individual. But according to inaccessible-essentialism, at most one of those individuals would be the real Holmes, to whom we refer in using the name, for given that one is the real Holmes, the others must be distinct though very similar people who are called by the same name. Yet the choice of one real Holmes from among the multitude would be arbitrary; it does not seem that any of the realizers would be the real Holmes to the exclusion of the others.

[18] Robert Howell (1979, sec. 7) protests that specification by description is not the only alternative to introduction by ostension; but his positive hint of a *tertium quid* is so far very obscure. (It seems structurally similar to Lewis' [1983] method of perceiver-relative counterparts.)

[19] Currie (1986) makes the point in terms of Ramsey sentences, claiming that grasping a Ramsification of a story is "all there is to understanding" the story.

Coburn (1986) asks why, if Haecceitism holds for actuals, we cannot imagine a world qualitatively indistinguishable from @ but containing entirely different individuals *and in addition* a second qualitatively indistinguishable world containing still a new, third set of distinct individuals (which seems absurd). The answer is that if the Conservative Position is correct, this absurdity is just what we should expect, but if Haecceitism regarding nonactuals should turn out to be true, Coburn's hypothesis is not absurd after all.

[20] Kraut (1979, p. 213) notes the peculiarity of someone's (truly) claiming to want *Pegasus* but denying that he would be satisfied by just any winged horse that was ridden by Bellerophon etc. etc.

[21] Note the ambiguity of "unexemplified" here as between "not instantiated at @," "not instantiated at such-and-such a world in which it exists," and "not instantiated at any world." In the absence of qualification I shall always mean the first of these.

[22] This final objection was put to me by Greg Currie.

[23] Lewis (1983, pp. 22-23; 1986, sec. 4.2) poses a seventh: that he can make no sense of the idea of an object's existing *in its own person* at each of two different worlds and nonetheless having properties at one that it does not have at the other. (He rejects the analogy of a thing's having a property at one *time* but not at another.) This intuition of bad craziness is of course the same as that which has always driven Lewis to counterpart theory as opposed to literal transworld identity; and it enables him to return at least some of the "incredulous stares" he is accustomed to receiving from complacent actualists. For my part, as we saw in Chapters 2 and 4, I have no difficulty at all with the notion of having properties relative to worlds; I find the time analogy entirely supportive here, even if in other respects it is imperfect.

[24] For McMichael's own semantics see again (1983b).

[25] A similar suggestion was made by Carney (1977). Forbes (1986) takes this idea seriously, but does not end up a Haecceitist regarding nonactuals; he favors an analogue of the Searle-Wiggins-Armstrong view I criticized in Chapter 5. I think this is in part because his conception of a fictional being hovers somewhere between van Inwagen's notion of a fictional character and what I have called a fictional individual or person as opposed to a character.

[26] Castañeda (1979) discusses an interesting actual case of this type.

[27] Obviously there are vexing borderline cases here. I am inclined *not* to treat pastiches in the way described; the character called "B*nd" in Christopher Cerf's and Michael Frith's Ian Fleming parody, *Alligator* (*Harvard Lampoon*, 1962), is not James Bond even though he is directly inspired by Fleming's creation. For that matter I do not countenance the protagonist of John Gardner's recent, comparatively respectful "Bond" books, for the reason that Gardner's hero is very distinctly Hollywood Bond--just good fun--as opposed to Fleming's own very serious character. (Mr. Fleming presciently made this distinction himself, in *The Man with the Golden Gun*, when Bond, amnesic, is being vetted for authenticity by members of his own Service: his suspiciously new clothing and his choice of the Ritz hotel, both in fact dictated by his KGB brainwashers, are described by the Chief of Staff as "sort of stage Bond.") I am more amenable to the Bond of Kingsley Amis (in *Colonel Sun*, written under the pseudonym "Robert Markham"), who seems totally genuine save for the interjection of one or two unmistakable and presumably irrepressible Amisisms. The difference in identity between Gardner's and Amis'/Fleming's character is due, I surmise, to differences in the authors' respective metalinguistic intentions, but the latter would be very hard to spell out.

[28] It may seem that my own case (presented earlier) of the absence of the editor figure and the presence of an entirely different character who happens to be named "Perry White" is a counterexample, since the distinctness of the two seems to consist in the difference between their descriptive roles in the story. This again depends on Mr. Siegel's intentions. If he had never created the editor figure in the first place, his alternative invention of the janitor or itinerant bum would have been simply a different fictive act, and so would have had a different character as its issue in any case. If, on the other hand, Siegel had written in the editor figure to begin with but (for some peculiar reason) then decided that White would not have been like that but would have pursued a very different sort of life-plan eventuating in a janitor's job or in vagrancy, the new White *would* count as the same character as the old despite his drastic transformation.

[29] As McMichael has pointed out in conversation, this requires that we conceive of fictional characters and worlds as being created by authors rather than as having existed from time immemorial. That is all right with me (especially since I have no great stake in the Haecceitist view I am now adumbrating); the view takes fictional characters to be something like fictive acts with qualitative stereotypes appended.

[30] Fine (1982) relies heavily on the author's "sayso" in the individuation of fictional characters.

[31] I am not so sure about Robert Coburn's (1986) question of whether God could have had a choice as between implementing Alvin Plantinga's haecceity and a putatively distinct haecceity ($\lambda x[x = $ Plantinga*]") that is not qualitatively distinguished from Plantinga's. Nasty theological questions and questions of Divine reference-fixing mingle here. I suspect, on behalf of my developing Causal-Historical theory, that our current use of the name "Plantinga*" fails to connect in the right way with any bonafide *dubbing* on God's part; but the matter needs considerable investigation.

[32] All one can say is: They are properties; they are nonqualitative; they are unexemplified; like the haecceities of actuals they are what track individuals through radical change from world to world; and their epistemology too goes by causal-historical chains. Do they indeed exist? Well, at a world distinct from ours, something has the property of being Holmes, and nothing has that property here.

[33] It is not clear whether we should count qualitatively identical double-Doyle worlds as distinct if in one "Holmes" #1 is (stipulated to be) the real Holmes while in the other "Holmes" #2 is. That depends on how we have antecedently distinguished #1 from #2, independently of the question of Homes' haecceity; I suspect there is no ultimately coherent way of doing so.

[34] That is not strictly right, as Currie has also observed to me; if one of the two planets contains actual individuals, such as Baker Street or Afghanistan, then that planet would be the one described by Conan Doyle. To secure the point we would have to take as our example an *entirely* fictional, fantastical work that made no reference to any actual item.

[35] McMichael takes as primitive an accessibility relation among maximally consistent "roles."

[36] This is Adams' (1981) view.

[37] Edelberg himself uses the example to motivate working with theory-immanent notions of truth and reference rather than the standard "realist" word-world notions.

CHAPTER 7

THE PARADOX OF NAMING RESOLVED, BY A KINDER, GENTLER THEORY OF DIRECT REFERENCE

There is overwhelming evidence that proper names must have senses or connotations that somehow contain contingent information about their referents. There is also overwhelming evidence that proper names cannot possibly have such senses or connotations. That is the Paradox of Naming.

The former, familiar evidence comes from Frege and Russell (and from their contemporary inheritors such as Schiffer [1978]) in the form of appeals to substitutivity failure, negative existentials, the informativeness of identity statements, and so on. The latter evidence comes largely from Ruth Marcus and Saul Kripke.[1] Kripke defended the opposing Millian tradition by pointing out in particular that if names abbreviated definite descriptions or otherwise expressed contingent properties of their referents, (a) certain sentences would be tautologous or contradictory that are not in fact tautologous or contradictory; (b) certain names that denote distinct individuals would denote the same individual; (c) serious problems would flow from the then presumed *equivocity* of names; and (d) a name would denote different individuals in different possible worlds.[2] Rather, Kripke maintains, names are rigid designators; a name denotes an individual quite independently of any contingent property the individual might have, and hence denotes that same individual in any world in which it exists. Mill seems to have been right in contending that (normally) the semantic function of a name is simply to pick out its bearer, and it follows that names are rigid in the sense defined.

I shall not here rehearse any of the well-known arguments on either side, but will try to sharpen the paradox in my own way.

136   CHAPTER SEVEN

1. THE PARADOX AND THE POSSIBILITIES

Suppose that on occasion O, S tokens "Cicero was bald, but Tully was not," in a suitably sincere tone, and goes on to behave accordingly. Observing this, we would all naturally want to infer

(A) S believes that Cicero was bald and Tully was not.

What possible positions might we take regarding the semantics, and resulting truth-value, of (A)? Let us assume for convenience that "Cicero" and "Tully" have the same semantical status, whatever that status might be; imagine an appropriate story leading up to O.

> (I)   *"Cicero" and "Tully" are connotationless, "Millian" names.* (As I have said, it is clear that all Millian names are rigid designators in Kripke's sense, but the converse does not hold. A Millian name is one whose *sole* semantical function is to pick out its bearer; contrast "the greatest prime smaller than 3,087," which due to the presumed necessity of arithmetic truths is rigid, but is not Millian.)
>
> > (a) S does not really believe what (A) says he believes; (A) is false.
> >
> > (b) S does really believe it; (A) is true and S accepts a contradiction,
> >
> > > (i)   Despite appearances, S is irrational.
> > >
> > > (ii)   Sometimes belief in contradictions does not reflect badly on one's rationality.

KINDER, GENTLER DIRECT REFERENCE 137

(II) "Cicero" and "Tully" are not Millian names; i.e., they make (distinct) semantical contributions of some specific kind, over and above denoting their common bearer.

    (a) They abbreviate flaccid (nonrigid) definite descriptions in (A), and perhaps also for S on O.

    (b) They abbreviate *world-indexed* or otherwise rigidified descriptions, and thus are rigid though not Millian. (This is Alvin Plantinga's "Boethian Compromise," more on which below.)

    (c) ...?

Option (I) is hard to maintain in the face of Russellian arguments. Indeed, Kripke himself impugned it in propounding his "Puzzle about Belief."[3] I think we can take (I) a good deal farther than many people suppose,[4] especially if we take suboption (b-ii) seriously (I shall be returning to (b-ii)), but it is not intuitively a very appealing view and it has some very nasty consequences (see Baker [1983]). Our only very compelling motivation *for* (I) is our desire to avoid (II-a), the Russellian option ostensibly discredited by Kripke's original criticisms. What we really want is some *tertium quid*, some hybrid or conciliatory synthesis somewhere and somehow in between Millianism and Russellianism. But what form would it take?

I propose the following constraint, which will help to bring out the difficulty of our dilemma.

THE RULE OF EFFABILITY: Any semantical account of names that constitutes a *tertium quid* of the sort we are seeking must be representable as the assignment of a logical form to (A), which logical form is expressed in standard logic plus a univocal belief operator. [Qualifications: (1) No ontological holds are barred; logical forms may involve

> quantification over any entities one likes. (2) Logical representations need bear no discoverable grammatical relations to (A).]

My justification for this rule is substantially just that I am hard put to think of any alternative format for a semantical theory. (Notice that the Rule does not forbid the use of intensional idioms; intensionality will come in, or may come in, by way of the introduction of intensional *entities* such as properties, propositions, possible worlds, or world lines. There is no fancy logic, however *outré*, whose semantic interpretation cannot be written in standard logic.) Now, the Paradox of Naming is generated by the assumption that our only possible options *anent* sentences like (A) are (I), (II-a), or something effectively equivalent to (II-a)--in short, the assumption that any name that is not Millian in effect abbreviates a description. I am inclined to think that what gives this (usually unexamined) assumption its plausibility is the Rule of Effability or something like it. If a name is not a semantically simple, unstructured individual constant at the level of logical form, then it must be some combination of constants, variables and predicates; the only type of combination that seems at all appropriate is a Russellized description.[5]

A word about "abbreviation": One excellent reason for denying hotly that names abbreviate descriptions is that it is fanciful at best to suppose that when an actual speaker tokens a sentence such as (A), an actual, psychologically real abbreviative *operation* has occurred within that speaker; surely nowhere in the speaker's mind has some Russellian quantificational structure triggered an abbreviating transformation that deletes that fairly rich structure and substitutes the tag "Cicero" or whatever. (On this robust account, different quantificational structures within different speakers, or even within the same speaker at different times, would just *happen* to be scrunched up and lexicalized into the same short word.) So it may seem obvious that if *this* is what is meant by "abbreviate" in (II-a), there must be a more plausible version of option (II)--one which takes "abbreviate" less literally--and so there is no paradox yet.

Quite so. But Kripke's objections to option (II-a) do not turn on the issue of the psychological reality of abbreviation; they are directed against even the much weaker suggestion that a proper name in an ordinary speech context is equivalent in *any* semantical way to any particular flaccid description or body of contingently descriptive material. So let us hereafter understand (II-a) in this weaker, nonrealistic sense of "abbreviate." I shall return to the issue of abbreviation in section 6 below.

2. PLANTINGA'S COMPROMISE

Plantinga's (1978) "Boethian Compromise" actually provides a clever and ingenious *tertium quid*, (II-b), that satisfies the Rule of Effability. It capitalizes on the previously neglected distinction between being Millian and merely being rigid. That distinction had always been illustrated mainly by throwaway examples from arithmetic that turn on the (alleged) necessity of all arithmetic truth, and so it never seemed particularly important. What Plantinga noticed is that for any contingent identifying property that any ordinary thing has, there is a corresponding "world-indexed" property that is an essential property of that thing, and for each such world-indexed property there will be an identifying description of the thing that is rigid but (because it has descriptive content and refers only by exploiting that content) not Millian. (Example: To "the winner of the 1962 Outboard-Motor-Eating Contest," a flaccid description, there corresponds "the winner of the 1962 Outboard-Motor-Eating Contest at @," which is rigid; at a given world w, it picks out, not whatever denizen of w wins the contest at w, but Hermina Schwartzenegger, the person who actually won the contest here at @.) Thus Plantinga seems to slip neatly between the horns of our dilemma.

Unfortunately, Plantinga is as firmly committed to the contradictoriness of our friend S's belief as is the Millian. Suppose "Cicero" and "Tully" abbreviate distinct world-indexed descriptions, both satisfied by their common referent. Both descriptions are rigid; that is what world-indexing is for. But if both descriptions are rigid, then S's utterance on O is true in no possible world. It refers doubly

140   CHAPTER SEVEN

to one and the same individual at each world and predicates contradictory properties of him. The proponent of (II-b) has just the same options *vis-á-vis* *S*'s rationality as has the proponent of (I-b). In this sense, the mere rigidity of names is just as bad as full-scale Millianness.[6] The same point can be brought out by a slight reformulation of our original dilemma:[7] As before, (A) seems to be true. But if names are rigid, then (B) *S*'s sentence is false at any possible world, and to believe what it expresses is to believe an explicit contradiction. Yet we may suppose that *S* is fully rational and would not accept any explicit contradiction; (C) *S* is committing no *fallacy* or crass logical error, however ignorant *S* may be.

3. THE PLANTINGA PROBLEM GENERALIZED

Of course, there are plenty of different kinds of semantical theories all directed towards problems of referential opacity and related matters. So it may seem surprising that anyone should think we had exhausted all our options so soon. What about inscriptionalism? Frege's own shift-of-reference view? Hintikka's theory of world-lines? All the solutions that Quine considers in his various writings on this topic? Etc., etc.

Some of these options, such as naïve inscriptionalism, are plainly inadequate from the beginning. More commonly, though, a view will be seen to *collapse into the Russellian option* even though the view seems more elaborate and mobilizes sophisticated apparatus. One such view is Frege's, as I read him. He hypothesizes that inside a (single) belief operator, a name shifts its reference from its customary bearer to its customary *sense*, an abstract constituent of the "thought" or proposition named by the complement clause as a whole; what causes substitutivity failure is that distinct names normally have distinct customary senses even though they may happen to have the same customary referents. But in Frege's examples,[8] the senses of names are expressed in the form of definite descriptions, implying that names *have* the senses of descriptions. Thus Frege seems committed to (II-a) and open to Kripke's objections.[9]

A more interesting example of a sophisticated theory that seems to collapse into Russellianism is Hintikka's classic possible-worlds version of Frege's approach ([1969] and elsewhere). Hintikka holds that inside belief contexts names name, not the ordinary inhabitants of our world, but "individuating functions" or world-lines. Each world-line, corresponding roughly though not exactly to a Fregean individual concept, is a function from worlds to individual inhabitants of those worlds. (On Hintikka's view, actually, it is the world-line itself that corresponds to a real individual, since real individuals such as you and I come complete with counterfactual properties as well as the actual properties we have here at @. This leaves us with the uncomfortable question of what sorts of things the values of world-lines are. Hintikka's followers sometimes call them "individual *manifestations at*" worlds; or we might call them world-*slices* of individuals.)

Each world line, associated with a proper name N, is determined or generated by a particular "method of cross-identification" or reidentification procedure governing the correct use of N (i.e., with a rule for answering questions of the form, "Which denizen of *w* is (identical with) N?"). Different names of the same person can be associated with different methods of cross-identification in the minds of believers; the procedure for recognizing Clark Kent in a given world, in the mind of a believer who is unaware that Kent is Superman, is obviously very different from the same believer's procedure for recognizing Superman at a given world. The reason that names do not substitute in belief contexts is that in those contexts they name, not the same earthly person, but distinct (because diverging) world-lines.

Notice that by virtue of his appeal to the reidentification procedures speakers actually use, which are "accessible" in the sense of Chapter 5, Hintikka is committed to the contingency of identity statements involving names and hence to the flaccidity of names, at least in belief contexts. (He remarks that "...even an allegedly rigid designator can be bent by epistemic considerations" [1975, p. xi].) World-lines are generated by sets of reidentification criteria that human believers actually use in tracing other things and people, or so we are left to suppose. But any such criterion must be substantive and useful as an epistemic tool; as Russell kept saying in "The Philosophy of Logical Atomism" (1918-19/1956), we do not recognize our friends by getting a look at their metaphysical insides and ogling their haecceities. We spot them by their appearances, their actions, their personalities, and so on--almost invariably, by clearly contingent properties they have. Thus, on Hintikka's view, any name will apply at world w to whomever or whatever has such-and-such contingent properties at w, rather than to whomever or whatever is the individual that the name intuitively denotes here at @. And this is just a version of thesis (II-a), the Russellian option; Kripke's original objections apply.[10]

It is all very discouraging.[11] But I think there is a genuine way out. It will take us at least briefly into philosophy of mind, because I think any real solution to the Paradox of Naming and/or to Kripke's Puzzle requires some attention to psychological matters.[12]

4. A PARATACTIC APPROACH TO ATTITUDE CONTEXTS

Wilfrid Sellars and a number of subsequent authors have defended the view that the sentential complement of a belief-ascription serves as a sort of *exemplar* or sample of what is said to be believed (Sellars [1956], [1963], [1967], [1969], [1973]; see also Davidson [1968] and Hill [1976]). For example,

(1) S believes that broccoli causes flat feet

is to be understood as having the structure,

(2) S believes some ·Broccoli causes flat feet·,

the dot quotes serving both to form what is grammatically a common noun and to ostend or demonstrate the linguistic token that they enclose. (Strictly speaking, the ostension is deferred, as we shall see.) Alternatively, perhaps a clearer representation of (1) would be

(3) S believes one of *those*.
↓

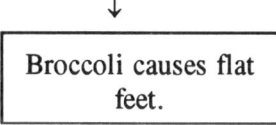

Belief, then, is a two-place relation holding between a person and an inscriptional token that falls under a certain type; a person has a belief when he or she bears that relation to such a token.

This grammatical/semantic account of belief ascription also makes a nice ontological/psychological theory of believing (or so I have argued (Lycan 1988]), one that fits nicely into a Functionalist theory of mind of the sort I favor. For there is good reason to think that people sometimes host inner episodes, which I construe as brain episodes, that have the sorts of structure that we associate with bits of public language (Sellars [1956], Harman [1973], Fodor [1975, 1978], Field [1978]). E.g., a single such episode might have subject-predicate form, i.e., might consist of the tokening of a mental representation of an object X concatenated with the tokening of a representation that expresses the property of being F; in so doing the episode would count as a ·X is F·, just as would a public utterance.

But in virtue of what would a brain episode or its parts be a *representation* of X or of F-ness or of anything else? How is it possible for purely physical brain events to have "aboutness" or intentionality? That depends on what we think it takes for anything physical to have aboutness, to refer or be intentional, and on whether our answer to that question is extrapolable from paradigm cases in public language, say to the more recondite case of brain activity.

This is now an old story, first told in modern times by Sellars and subsequently amplified and sophisticated by Harman, Fodor, Field, and many others. It directly involves three questions: (a) What does it take for some brain item to have the intentional content *that P*? (b) More ambitiously, what does it take for the item to be a *representation* that P?[13] (c) Given that an item represents that P, what does it take for the item to be a *belief*, or any other particular propositional attitude, that P? (a) and (b) are the main questions of "psychosemantics" in the sense of Fodor (1987, 1990); (c) is provisionally answered by Functionalist theories of mind.

A lesson of the past twenty years' work in the theory of reference is that what mediates linguistic aboutness is causality--causal chains of certain roughly specifiable types,[14] and/or reliable functional or teleological links consisting in linguistic expressions' specific sensitivity to their referents (cf. Harman [1973], Bennett [1976], Stampe [1977], Stalnaker [1984], Millikan [1984], Fodor [1990], and many others). However the details of these connections might go, the important points to note are (i) that the connections are *naturalistic* (they are real relations to be found in nature), and (ii) that other physical items besides public linguistic tokens can bear them to objects--brain events, in particular, can have the appropriate sorts of etiologies and can be innate or learned responses showing specific sensitivity to particular objects and types of object. Thus, we have every reason to think, or at least to hope, that brain events can be intentional in (almost) just the way that bits of language are.

So: To judge or believe that P[15] is to bear the "belief" relation to an inner representation whose syntactic/semantic structure is analogous to that of the sentence that replaces "P" (more on which analogousness shortly). The "belief" relation itself is a distinctive *functional* relation, consisting in the representation's playing a certain type of functional role, i.e., doing a certain type of administrative job within the functional hierarchy that is the believer him- or herself. Obviously the role that is characteristic of beliefs as opposed to desires, intentions, and so on has to do with storage, with mapping, and otherwise with serving as a guide to action.[16]

(There is a great tendency in the literature[17] to overstate the commitments of this representationalist theory, and indeed to caricature it unmercifully, as if the representationalist were suggesting that inside each believer's head is a tiny blackboard with all the believer's stored representations written in chalk, or that evil, politically motivated scientists might be able to spy inside our heads with their cerebroscopes and report our innermost convictions to the Thought Police. I have disavowed these straw-man interpretations in previous works (see particularly Lycan [1988, 1990, 1991], and tried to show why no such absurd consequences in fact flow from the representationalist view, so I shall continue to assume the representationalist picture here.)

So far none of what I have said about intentionality is at all original. But now it is time to unveil the gimmick I shall use in attacking the Paradox of Naming. Let us turn back to the mechanics of dot quotation.

5. A TWO-SCHEME THEORY

Dot quotes, and plural demonstratives of the sort that occur in (3) above, *classify* the tokens they ostend. Now, how do they do that? How do we determine the extension of "...is a ·Broccoli causes flat feet·," or tell when some token of a very different shape *is* "one of *those*"? Philosophers have offered varying advice about this. Sellars himself counts an item as a ·Broccoli causes flat feet· just in case it plays approximately the same *inferential role* (augmented by "language-entry" and "language-exit" norms) within its own surrounding conceptual framework as the English sentence "Broccoli causes flat feet" plays within English. Davidson merely introduces a relation of "samesaying" without further comment, but given the semantical format for which he is famous, we might suppose he would count something as a ·Broccoli causes flat feet· just in case it has the same truth condition as does the corresponding English sentence; or we might insist on a stronger intensional isomorphism, perhaps based on the procedure by which the truth-condition is computed. Other individuative methods are possible also, and the multiplicity of possibilities is what is going to help us resolve our paradox.

Let us ask ourselves, as Kripke (1979) does by implication: Just what does poor S believe about Cicero and about Tully? Does he believe that Cicero was bald? Less tractably, *does he believe that Tully was bald*? There seems no obstacle to answering "Yes" to the first of these questions, but intuitions divide sharply over the second. Impressed with the felt rigidity of names, I used to maintain (e.g., in Boër and Lycan [1980]) that S just plain does believe that Tully (that very person, whom S himself happens to be calling "Cicero") was bald, even though *S* himself would never express his belief in that way, and that *S* does hold contradictory beliefs, though he has no way of detecting this. (He also believes that Cicero (=Tully) was not bald; another contradiction.) Other philosophers are more impressed by the commonsensical argument that if *S* sincerely, coherently, and lucidly asserts "Tully was not bald," then *S* believes that Tully was not bald (and doesn't believe that Tully is).

An irenic solution is available. Both intuitions should be respected, and thanks to the Sellarsian semantics they can be, in a way that has independent motivation as well. Recall that we have a choice of classificatory schemes to impose on our proffered exemplars, i.e., of methods for telling which tokens count as ·[so-and-so]·'s and which do not. Proponents of the Sellarsian approach have always tended to suppose that some one such scheme is the correct scheme. But suppose we reject that presumption instead. There is considerable independent evidence[18] that at least two disparate individuative schemes are used on different sorts of ascription-occasions, depending on the context and the conversational point of the ascription. Sometimes we follow Sellars himself in classifying our tokens according to their inferential roles, or perhaps I should say, according to their *computational* roles *á la* Jerry Fodor,[19] when what interests us are the tokens' causal properties and the explanation and prediction of behavior. (The lesson of "methodological solipsism" as I understand it is that a belief's computational surface and (emphatically) not its truth-condition or other semantical property is what determines its causal contribution strictly construed.) But sometimes we do individuate beliefs semantically, e.g., according to sameness of truth-condition, when we are more interested in truth-related aspects of belief, such as the believer's success in achieving his/her goals, or the success of our use of his/her belief reports as authoritative evidence for our own beliefs.

Our two-scheme hypothesis affords us a properly equivocal resolution of our dilemma and a likewise equivocal answer to Kripke's type of hard question. First, does *S* irrationally believe a contradiction? When he makes his celebrated utterance, he does so partly as a result of having affirmed a mental version of the sentence he utters; he has accepted a ·Cicero was bald and Tully was not·. The

Millian, or even the proponent of names' rigidity, will insist that this amounts to accepting a ·Cicero was bald and Cicero was not·, since if names are Millian or at least rigid, the names "Cicero" and "Tully" make exactly the same truth-conditional contribution, and the two tokens just displayed between dot quotes express exactly the same belief in the sense of being true at just the same worlds (and for the same reason). But is S's mental analogue of "Cicero was bald and Tully was not" really a ·Cicero was bald and Cicero was not·? That depends on our choice of individuative schemes for dot-quoted items. If we appeal to the semantical scheme, we find it does count S's mental token as a ·Cicero was bald and Cicero was not·, and in that sense S does believe a contradiction. But *option (I-b-ii) is in force--in the present sense*, there is nothing irrational about believing a contradiction. For S has no syntactic way of detecting his semantical anomaly; he cannot deduce from his mental token anything he could recognize as a contradiction. The relevant contents of his head are analogous to an uninterpreted formal calculus equipped with rules of natural deduction. In the absence of "a = b" as an axiom, it would be positively irrational to infer by substitution of "b" for "a" even if on some preferred interpretation that *could* be supplied by an external observer, "a" and "b" are assigned the same referent.

Now, what about appeal to the computational individuative scheme? Computationally speaking, S's mental token of "Cicero was bald and Tully was not" does *not* count as a ·Cicero was bald and Cicero was not·, since the representations associated respectively with the names "Cicero" and "Tully" play obviously distinct inferential and computational roles for S, and accordingly distinct behavior-causing roles.[20] From "Cicero was bald and Tully was not," e.g., S would infer "Cicero was bald," "Tully had a property that Cicero didn't," "Cicero and Tully were two different people," etc., and would *not* infer "Tully was bald" or "Tully existed." But from "Cicero was bald and Cicero was not," S would either start inferring every sentence he could think of, or go into cognitive spasm of some sort (given a generous helping of downward causation, S's circuitry might turn black and give off smoke). Thus, in this sense (according to this individuative scheme), S does *not* believe that Cicero was bald and

Cicero was not, even though he believes that Cicero was bald and Tully was not. It is probably psychologically impossible to believe an explicit contradiction *computationally individuated*.

Several authors, early on, glanced off the distinction between our two individuative schemes, without entirely realizing that that was what they were doing. Hartry Field, Jerry Fodor, David Lewis, and Stephen Stich hinted at it.[21] But, interestingly, they have in effect taken sides (different sides) on which of the two schemes is correct, or at least on which is vital to the concept of belief and which negligible. Fodor and Lewis assumed that beliefs are essentially causal entities invoked to explain behavior and that their semantical properties are by the way, while Stich and Perry insisted that the truth-values of beliefs (and the reliability of informants) are what matter and that explaining behavior is not so important after all. Now, this seems to me a funny sort of thing to *quarrel* about. Sometimes we are interested in explaining and predicting behavior; at other times we are interested in truth and reliability. Which of these interests is *objectively paramount* seems to me an idle question. And if my Sellarsian semantics is right, our language affords us a pragmatic choice in belief ascription, that matches our pair of alternative interests nicely.

More recently yet, theories have been put forward that incorporate *both* my sorts of individuation scheme and treat belief content as a pairing of causal role and truth condition.[22] It is important to see that my own view does not amalgamate the two schemes in that way, but keeps them alternative, with a resulting pragmatic ambiguity in attitude ascriptions.

Does S believe that Tully was bald? On our two-scheme hypothesis the answer is, quite properly, "Yes and no." On the computational scheme, S's mental analogue of "Cicero was bald" does not count as a ·Tully was bald·; S's mental "Cicero" and his mental "Tully" play entirely different computational roles, even though they are in fact grounded in (are representations of) one and the same person. On the semantical scheme, S does believe that Tully was bald, since that scheme does count his inner analogue of "Cicero was bald" as a Tully was bald. Thus, by providing for an ambiguity in the reference of the plural demonstrative underlying the complementizer

KINDER, GENTLER DIRECT REFERENCE 149

"that," our Sellarsian account is able to predict and explain our preanalytic uncertainty and disagreement about "*S* believes that Tully was bald."[23]

6. ABBREVIATION AND THE SPOT-CHECK TEST

Let us briefly revisit the topic of *abbreviation*, for it too is illuminated by our two-scheme hypothesis. In section 1 we rejected as preposterous the suggestion that an abbreviative procedure literally occurs whereby Russellized descriptions are squished into names. We have also wanted to reject (II-a), the Russellian option, *tout court*, and to deny that names are even *equivalent* to flaccid descriptions in any semantical sense. Yet something funds the Russellian intuition, and that something is brought out by what we might call the "spot-check test." (I use this in teaching Russell to undergraduates.) Suppose we overhear Jones using a certain name, say, "Wilfrid Sellars," and we want to test his grasp of his own idiolect. Whipping around, we demand, "Whom do you mean by 'Wilfrid Sellars'? Quick, now!" Subjected to this test, Jones will doubtless produce a description, as in "Oh, I mean the late famous Pittsburgh philosopher who wrote 'Empiricism and the Philosophy of Mind'." This phenomenon suggests that the ability to produce a description on demand is constitutive of competence in the use of a name, and Russellians might take it as proving the synonymy of names with descriptions. (That it does *not* prove that is what Kripke showed in enforcing his now well-known distinction between a description's actually "fixing the sense" of a name and the description's merely "fixing [the name's] reference.") But now we are in a position to explain the results of the spot-check test without incurring the consequence that (II-a) is correct.

A Mentalese name, as used by a person at a time, can have the same *computational role as* a description for the person at that time. Thus it can be "equivalent to" the description *modulo* the computational individuative scheme and two of the subject's beliefs, alike except that one involves the name while the other involves the description, will be computationally and hence causally similar for the subject at the time-- the two beliefs will be *functionally equivalent* for all practical purposes without being semantically equivalent at all. Why should names ever share the functional roles of descriptions in this way? I think the answer must be that functionally speaking, names are something like labels on files, or perhaps more like *tabs* on files, where each file is a store of contingent information associated with the name. A tokening of the inner name calls up the most salient information in the file (perhaps tokening just *is* the calling up of that information), and that is why we feel that particular uses of names are "backed by" particular bodies of descriptive material. It is also why the spot-check test works. (It is also, I shall argue, why identity statements involving names are "informative" despite the triviality of their semantical contents: When an identity statement is accepted, files get merged;[24] and cognitive capacities thereby usefully consolidated. More on this shortly.)

Names do not abbreviate descriptions in any semantical sense at all. They just share computational roles with descriptions from time to time.

I argued in Lycan (1988, Ch. 4) that the two-scheme hypothesis also solves the problem of self-regarding attitudes, succeeds in sorting out some puzzle cases, and cures misconceptions about methodological solipsism. I will not repeat those discussions here,[25] but instead compare my view to current "Direct Reference" theories of names (see fn 4). It is very similar but, I shall argue, superior.

7. DIRECT REFERENCE

I have always tended to think of myself as a "Direct Reference" theorist. But Devitt (1990) has convinced me that I am not one, in the label's fullest sense. A *real* DR theorist simply does not admit substitution failure for names in belief contexts, period, but tries to explain away the appearance of substitution failure; whereas I do admit, indeed predict, genuine substitution failure modulo my computational/functional scheme. But Devitt has not convinced me that I have "held onto the rhetoric of DR but squandered the substance" (as he put it in discussion), and the reason is interesting.

Technically I agree that (on my computational/ functional scheme) the names make different contributions to the containing sentence's truth-condition, and that is why I am not an absolutely pure DR theorist. (Hereafter I shall speak of "Doctrinaire DR."[26]) But they make their differential contributions only in a very peculiar way--not in virtue of what I claim are their semantic properties, but in virtue of the extraneous mechanics of the dot quotes. It is a special effect, very like the one that figures in Quine's (1953/1963, pp. 139-40) famous "Giorgione" example:

(G) Giorgione was so-called because of his size.

One cannot substitute the coreferring name "Barbarelli" for "Giorgione." But the failure owes, not to any semantical property of the name "Giorgione," which (for all the example shows) simply and directly refers to Barbarelli, but to the fact that an entirely separate element of sentence (G)--the morpheme underlying "so"-- metalinguistically denotes the name "Giorgione" itself; the latter element loses its referent when "Barbarelli" is substituted into (G), and that is why the truth-condition of the resulting sentence differs. Thus, the following nearly irresistible Quinean principle turns out to be false: "If substitution fails, then at least one of the names in question must be contributing more than just its bearer to the truth-condition of the containing sentence."

So: Doctrinaire DR fails in (G), because substitution does. But substitution fails, not because the two names' semantics differ in the slightest, but because the metalinguistic demonstrative in "so-called" introduces reference to the particular name occupying the slot. Semantically, the name itself just refers to its bearer; but its own semantics is, oddly, not its only contribution to the semantics of the whole sentence, because it happens itself to serve as the referent of an entirely distinct term elsewhere in the sentence.

A similar thing happens in belief ascriptions, on the Sellarsian view. Their complements are really displayed tokens, produced with no illocutionary force. The names occurring in them have their normal referents and their normal semantics, i.e., simply designating their bearers, just as DR has always insisted; but the extraneous dot quotes are metalinguistic, and so, when the computational scheme is in force, they induce a change of extension and accordingly a change of truth-condition for the containing belief ascription when a substitution is made in the name slot. As in the "Giorgione" case, substitutivity does fail, but not because the names themselves do anything semantically but designate their bearers.[27]

Thus an unexpected distinction: Contra Doctrinaire DR, it is true that the names contribute more than simply their referents to belief ascriptions in which they occur. But it does not follow that semantically they do more than what Mill said they did. What they additionally contribute to belief ascriptions is not their own semantics,[28] but service as partial referents of other elements occurring in those sentences, the belief operators' complementizers. And that is why I have not merely borrowed DR's rhetoric, but held fast to its substance, while at the same time going Doctrinaire DR one considerably better by admitting the genuine failure of substitutivity. (But for Devitt [1990], I would not have seen this splendid feature of the Sellarsian view.)

Upshot: I see no reason to accept Doctrinaire DR, but belief contexts have given us no reason to abandon DR as a thesis about the semantics of names. We can and should preserve a kinder, gentler DR (hereafter Kinder, Gentler DR).

8. FREGE'S PUZZLE

If ordinary proper names are purely referential, as is claimed by Kinder, Gentler DR as well as the extremist variety, why are identity statements involving names so often informative? On either version of DR, any two such identity statements directed upon the same individual have the same truth-condition, no matter how finely "truth-conditions" are individuated, and express the same proposition by anyone's standards.

As everyone knows by now (Kripke [1979], Kaplan [1990]), the standard sort of Fregean example, say "Cicero = Cicero" vs. "Cicero = Tully," is somewhat misleading. Orthographic sameness of name does not guarantee non-informativeness ("Paderewski [the pianist] = Paderewski [the politician]"), and orthographic difference does not guarantee informativeness, depending on audience ("Descartes = René Descartes). In one way, this point is a red herring, for given DR, the mere existence of informative identities is still a puzzle, whether or not informativeness lines up with identity-sentence-types counted orthographically. But the point does indicate something of crucial importance: that "informativeness" is on its face relative to audience. Any identity statement considered merely as a sign-design could inform some suitably deprived audience, and no identity statement is guaranteed to inform every audience.

In that sense, no identity statement is "informative" or "uninformative" on its own. An identity statement informs, or fails to inform, in a context, depending on what its audience already knows and on what modes of presentation the audience is (psychologically, not semantically) associating with its component names.[29] And this is Kinder, Gentler DR's key to Frege's Puzzle. To inform is to affect someone's cognitive state, paradigmatically that person's beliefs.

If S is a person to whom an identity statement $\ulcorner N_1 = N_2 \urcorner$ does *not* come as news, S must already have a single file bearing counterparts of both N_1 and N_2 as labels. S's internal "N_1" and "N_2" are already functionally interchangeable, and so S's cognitive state is not significantly affected by the encounter. But if, antecedently, S's "N_1" and "N_2" attached to separate files, accepting the new identity statement effects a merger, and major functional events ensue. Substitution inferences are licensed, as are conjunctions of material previously separated across files, and then further consequences of such conjunctions that did not flow from either conjunct alone. Thus will the merger[30] have significant cognitive and behavioral effects--despite the semantic (truth-conditional) triviality of the instigating identity statement itself.

It is perhaps too easy to gesture toward "significant cognitive and behavioral effects" and let Frege's Puzzle go at that. The Puzzle was to show, more specifically, why someone who is "informed" by an identity statement *learns something*, and *what* the person learns. The "something" seems to be propositional in nature, and whatever proposition it is seems to be a nontrivial one. I have not spoken directly to that issue.

Suppose, then, that S is provided with authoritative testimony, "Cicero is Tully," and accepts it. Thus S now comes to believe that Cicero is Tully, ostensibly *not* having believed or realized it up till the revelation. But this brings us back to a now familiar sort of question: Did S not previously believe that Cicero was Tully? And on the two-scheme theory, that question requires disambiguation as always: (a) Did S believe a ·Cicero is Tully· semantically individuated? And (b) did S believe a ·Cicero is Tully· computationally individuated?

Assuming that S had previously heard of Tully at all, the answer to the first of those two questions is, unequivocally, "yes." Any ·Cicero is Cicero· or any ·Tully is Tully· is semantically speaking a ·Cicero is Tully· as well, and S doubtless accepted some of those. Thus, on my view, there is a sense in which S did believe all along that Cicero was Tully. I would argue that this is correct; there is such a sense, though it is not the dominant reading. If that seems odd, remember that every belief sentence whose complement contains a name *has* a referentially transparent reading, however recessive that reading is or misleading to use, and transparency is all that is required here.

The more interesting question is the second, and its answer is "no"; S did not believe a ·Cicero is Tully· computationally individuated. For part of "Cicero is Tully"'s inferential role--some logicians would say *the key* part of it--is to license substitution inferences, and prior to hearing "Cicero is Tully," S did not have that license in any form. In that sense, the computational sense, S has indeed learned something-- that Cicero is Tully--and hence been informed.[31]

This is an extremely important advantage of Kinder, Gentler DR over Doctrinaire DR. The Doctrinaire theorist cannot admit any sense in which S has newly learned *that Cicero is Tully*, but must fall back on the ancillary but distinct facts that S has also learned, such as that the two names corefer. I suspect that this implausible consequence more than any other feature of Doctrinaire DR has repelled readers who might otherwise have been persuaded by the arguments for DR.

It remains to explain why the computational reading is

dominant and the truth-conditional reading is hard to hear. One reason is that by their nature, identity statements play an unusual and distinctive informing role: (a) Their semantical content is both intrinsically uninteresting and normally not new to the subject; (b) their computational role (file-merging, with the ensuing flurry of inferential activity) is so dramatic, much more so than, say, the addition of a simple predication to an existing file; and (c) the behavioral effects of accepting a new identity statement flow from that computational activity rather than from a change of laconic truth-conditional content. These facts conspire to suppress interest in the semantical individuation scheme for attitude complements and enforce interest in the computational scheme.

A second reason is that, as DR theorists have pointed out at length,[32] an utterance of "*S* believes that Cicero is Tully" made before the revelation would be extremely misleading, even if true *and* even if flagged for semantical rather than computational individuation. Assuming the speaker knows that Cicero is Tully *and the semantical scheme is in force*, why on earth would the speaker switch names in mid-sentence? It is as if I, speaking of my daughter Katherine Jane Lycan (who goes by her middle name) were casually to say, "Katherine wonders whether Jane will sink as many foul shots tonight as KJL did on Saturday," instead of simply "Katherine [or Jane] wonders whether she will sink as many foul shots as she did on Saturday." But if the speaker does not know that Cicero is Tully (and the semantical scheme is still in force), the speaker would not be in a position to report either *S*'s belief that Cicero is Cicero or *S*'s belief that Tully is Tully by saying "*S* believes that Cicero is Tully" in the first place.

Moreover there is an indelible though weak presumption that even in a referentially transparent belief ascription, the words used in one's complement clause will not be unacceptable to the subject in question. For one thing, it is the charitable policy; when we are describing another person, it seems unsympathetic deliberately to describe him/her from an opposing point of view. For another thing, unless one is using a stilted explicitly *de re* construction ("Castro's island is such that 'way back then, Columbus believed it to be..."), the very availability of the computational individuative scheme will cause a hearer to waste a fraction of a second ruling out the computational reading. In fact, for this reason we have a *general* inclination toward computational readings of attitude complements; transparent readings are always a bit forced. For all these reasons, it is hard to hear the transparent reading of an identity attitude ascription.

Thus does Kinder, Gentler DR triumph by respecting the key datum, while of course preserving Doctrinaire DR's pragmatic style of explaining away the embarrassing consequence they share. It is better to admit a dominant reading on which S does newly learn that Cicero is Tully and apologize for the mere existence of a recessive reading on which S held that belief all along, than to deny the existence of the dominant reading and try to maintain that the recessive reading is the sole, univocal meaning of the target sentence.

9. EMPTY NAMES

Doctrinaire DR can tough out the problem of belief complements, albeit not very plausibly and, in light of Kinder, Gentler DR, unnecessarily. But Doctrinaire DR hits the wall at empty names. For precisely Russell's reasons, a "name" that has no referent is not genuinely a name in the sense of DR, and cannot simply contribute its bearer to the truth condition of a sentence in which it occurs.

However, on the face of things, Kinder, Gentler DR fares no better, for empty names lack referents outside propositional-attitude constructions as well as in. "Dupin was a detective," "Sherlock Holmes was a detective," "Poirot was a detective," and (if we suppose that posterity is badly mistaken in thinking there was a real Pinkerton) even "Pinkerton was a detective" would all have just the same degenerate truth condition--either *none*, or the same "truth condition" as has the open sentence "x was a detective," and they would all have the same meaning, ditto. This looks very bad.

Notice that the difficulty is truth-conditional, a semantical anomaly. It is not merely the related problem for philosophical theories of referring, faced most conspicuously by Causal-Historical theorists (e.g., Donnellan [1976], Devitt [1981]), of how best to explicate the notion of referring to a nonexistent. The latter problem is, though a problem, tractable; the former can seem simply prohibitive, and would remain unsolved even if an analysis like Donnellan's or Devitt's should prove entirely successful. (Incidentally, as is documented and deplored by Devitt [1989], there is a general tendency to confuse DR or Millianism in semantics with the Causal-Historical theory of referring in the philosophy of language. One must remember that the two are mutually independent and in fact have not been held by the same theorists.)

Surprisingly, or not surprisingly, little is to be found on empty names, in the existing DR literature.[33] What are the DR theorist's options?

The most obvious move would be brisk tactical withdrawal from DR. Nowhere is it written that a theorist who accepts DR for names of actual individuals must also accept DR for empty names. Indeed, a superficially plausible Russellian case can be made, along the lines of the previous chapter's Conservative Position, for a different treatment; in particular, since we cannot be perceptually acquainted with or even causally connected to a nonexistent, it seems we must know nonexistents only by description. But that sort of argument has been embarrassed by Kripke (1972) and his descendants such as Devitt (1981) between them; and we have already seen some difficulties that beset the Conservative Position, most notably McMichael's problem.[34]

A further objection to bifurcating the semantics of names is that often we do not know whether or not a particular name denotes. (Examples: "Arthur" as ostensibly referring to the ruler of Camelot; "Robin Hood"; "Donald" as allegedly denoting a particular CIA informant in Jim Garrison's investigation of the murder of President Kennedy.) Indeed, if we are fairly skeptical about historical figures whose names appear in no more than one ancient text, there are many quite familiar names whose referents are in doubt. For any such name, on the bifurcated view, its very semantics is in doubt as well: *If* the name does denote, it is a genuine, logically proper name and refers directly, but if its designatum is nonactual, it is not a name at all, but only speciously a "name," and really abbreviates a flaccid description or the like. That is an ugly feature of the bifurcation strategy. We are used to the idea that *lexical* meaning depends on contingent and a posteriori facts about nonlinguistic reality, but semantic and (surely) syntactic structure should not.

I do not regard those objections as conclusive, but they do motivate an afurcated approach, or at least an exploration of the oxymoronic idea of DR for empty names.

(i) Someone might be a DR Meinongian, and contend that "empty" names refer, directly, to Meinongian individuals.[35] But this would involve being a Meinongian, and so will not do for my purposes. (Also, one would need to provide an appropriate theory of the referring relation, one that did not posit semantic mediation, such as descriptive content, between name and Object, though the Causal-Historical theory sketched at the end of Chapter 6 might be mobilized here.)

The availability of a DR Meinongian position marks an interesting difference between Meinongianism and Mature Lewis. On Lewis' view, of course, there exist uncountable hosts of, e.g., Sherlock Holmes figures, in as many Doylean worlds (each figure named "Holmes" at its home world). Our name "Holmes," as we use it here at @, cannot very well designate any one of those otherworldly Holmes figures to the exclusion of the others; and as we saw in Chapter 2, Lewis cannot tolerate haecceities. Lewis' (1978) answer is that "Holmes" is always used flaccidly: "[T]he name denotes at w whichever inhabitant of w it is who there plays the role of Holmes" (p. 41). Thus Lewis seems forced to take the Conservative Position in regard to fictional names. And since he agrees that "ordinary" proper names are used rigidly, this makes him a bifurcation theorist; he is pushed away from DR.

Happily, we are not Lewisian Concretists in any case. Let us get on to the options available to the Ersatzer.

According to the Conservative Position considered in the previous chapter, nonactual individuals have "accessible" qualitative essences. That makes their names semantically equivalent to definite descriptions, since a name and its bearer's essential description will corefer at every world. Thus the Conservative Position does not afford a notion of direct reference. But what of my new Haecceitist proposal for nonactuals?

(ii) If we can make sense of the idea of a "blank" individuating property in the way I worked toward in Chapter 6, then distinct haecceities will yield distinct Ersatz singular propositions, and we could regard this as an Ersatz form of direct reference. (Though only Ersatz, because no flesh-and-blood individual *per se* figures in the "singular propositions" thus created, or in the truth conditions of sentences about the nonactual individuals in question.) The semantical notion of reference, as always a primitive within semantic theory, could again be philosophically cashed in terms of my nascent Causal-Historical theory.

However, as I said at the time, I am not entirely confident that the idea of a "blank" haecceity does make sense, or even that my Haecceitist proposal for nonactuals is sane. For that reason, I want now to offer an alternative DR treatment of empty names. That treatment will exhibit yet another advantage of Kinder, Gentler DR over Doctrinaire DR.

(iii) Let us take seriously the appalling suggestion that a sentence containing an empty name has either no truth condition or an incomplete one; such a sentence expresses no proposition, but at best a propositional function. If to have a determinate meaning, a sentence must have a truth condition and express a complete proposition, then sentences containing empty names do not have determinate meanings, and are in that sense *meaningless*, though they are still meaningful in the incomplete way that open sentences are. Frege and Strawson were quite right in maintaining that such sentences lack truth-value.

That harsh doctrine (call it the Harsh Doctrine) would apply to all fictional sentences, and to all sentences containing names that are in fact empty, no matter how well justified someone might be in thinking the empty names to be nonempty.

At first, the Harsh Doctrine sounds (besides preposterous) as arbitrary and *ad hoc* as the bifurcated view: An ordinary sentence has its truth condition in the normal way, determined in part by the (direct) referents of the names occurring within it; but a sentence containing an empty name is without a proper truth condition, meaningless. Thus when we are in doubt as to whether a name denotes, we must be in exactly as much doubt as to whether we have a meaningful sentence at all. Something has gone badly amiss. Brentano and Meinong maintained that a thought is the thought it is quite independently of the existence or nonexistence of its objects; so too, surely, a sentence is meaningful independently of the existence or nonexistence of its referents.

But no. There is a crucial difference between the Harsh Doctrine and the bifurcated view, in light of which the Doctrine is not nearly so arbitrary and implausible. For according to the bifurcated view, empty names have hidden semantics of their own, including hidden conceptual content (as always, the paradigm is semantical equivalence to a definite description). Moreover the complex semantics has to *be* hidden in a way somehow mediated by syntax; some device has to squash the complex content into what is superficially a simple label like "Holmes." The bifurcation theorist has to admit that until we are sure whether a particular name actually denotes, we are not sure whether all that apparatus is taking place behind the speech and understanding of users of that name. And, metaphysically, the existence of all that apparatus would depend on the contingent actual existence of a referent out in the world, perhaps millennia ago. The Harsh Doctrine carries none of this obnoxious baggage. For me, all that depends on the actual existence of a referent is whether the sentence in question is about something. If, at the relevant position, it is not about anything, then it is defective in meaning; so its semantic *value* depends on a contingent external state of affairs (no surprise there), and, granted, its expressing a complete proposition does so as well. But there is no semantic or syntactic *apparatus* whose existence and operation has to be hastily posited if we should discover that a particular name is empty.

All this distancing myself from the bifurcated view has merely postponed the more basic preposterousness question. How could anyone claim that sentences containing empty names are meaningless? Everyone understands a sentence like "Pegasus had wings," and what can be understood is meaningful. Moreover, as an imagined opponent said above, one's mere understanding of a sentence could hardly depend on what did or did not occur long ago and far away.

Beware that last principle, and for that matter beware the dictum of Brentano's and Meinong's that inspired it. If we are typing thoughts according to my computational/inferential scheme, the dictum seems true, but if we revert to the semantical individuation scheme, the dictum goes false. As has amply been pointed out in the "methodological solipsism" literature (see again fn 18), the "content" of a thought in the semantical, propositional or truth-conditional sense is not determined by what is in the head, but indeed partly a function of contingent external referents. And the same is true of sentence meanings. Thus no one should be surprised that the proposition expressed by a sentence should change depending on "external" states of affairs, or even that it should go downright defective if the world misbehaves badly enough.

The latter consideration undercuts the nascent argument I have ascribed to my opponent. But it does little else to salve the Harsh Doctrine's preposterousness. To improve the situation, let me offer some remarks that I think will help to make the Doctrine less awful.

First, consider two extremes. At one extreme, a name or "name" has no known history and no conventional associations at all, as in the following sentence:

(4) Blork is a detective.

"Blork" is (I suppose) functioning grammatically as a name, but it neither names anything nor has any other semantical feature--a *truly* empty name. In this case, I trust all would agree, there is a semantic hole. (4) is an only partially meaningful sentence, and does not express a complete proposition; the Harsh Doctrine is true of (4).

At the other extreme, a "name" simply does abbreviate a description. Any DR theorist should admit that there are such names, and perhaps that any proper name has possible flaccid *uses* (see Devitt [1974, p. 196], Boër [1978], Ziff [1978], Nunberg [1993]); DR's Millian claim concerns only the ordinary uses of most proper names. Names like "Chummy" (in British police novels) and "Podunk" (as meaning some American hick town) have descriptive meaning and denote, when they do denote, only in virtue of that meaning. Some mythic or legendary names arguably have fixed descriptive senses as well--e.g., "Santa Claus," which no longer denotes any historical person (if it ever did).

Thus we have two categories of empty names that do not trouble DR or the Harsh Doctrine, the first because it confirms the Doctrine and the second because DR does not apply to it (a *little* bifurcation is no terrible thing). The objections to the Harsh Doctrine come from cases in between our two extremes, that is, cases in which a name does have some historical context and/or conventional associations but is not semantically equivalent to any of that descriptive material. These include fictional names, if the Conservative Position is indeed mistaken, as well as legendary names that are erroneously thought to denote, and other names that we have heard somewhere and got the idea that they denote even though they do not. It is for these intermediate cases that the Harsh Doctrine seems most implausible. But here is where Kinder, Gentler DR is able to offer a palliative that Doctrinaire DR lacks.

I suggest that what convinces us of the perfect meaningfulness of sentences containing fictional etc. names is that such sentences are *cognized* in just the same ways as are fully denoting sentences. We can believe or disbelieve them, wonder about them, hope or wish them true, and so forth. It is this fact, I think, that constitutes our *understanding* of them, seemingly contrary to the Harsh Doctrine. And the same fact remains a massive obstacle for DR: If an empty name is entirely meaningless, it is meaningless inside attitude contexts as well as out; if "Poirot had a moustache" has exactly the semantics of "x had a moustache," then "S believes that Poirot had a moustache" has exactly the semantics of "S believes that x had a moustache," and expresses no complete proposition. But that is absurd.

Anyone who followed my analysis of Frege's Puzzle will know what is coming next. The key difference between Kinder, Gentler DR and Doctrinaire DR is that the former allows, as the latter does not, a "narrow," computational or inferential reading of attitude complements in addition to their semantical or truth-conditional readings. The computational/inferential habits of any relevant internal representations (a mental name-counterpart) will be generated by the descriptive material that *psychologically* "backs" the name. Thus, although there is an interpretation of the foregoing belief-ascription which is semantically defective (what S allegedly believes has no subject at all, and lacks truth-value), there is also an important sense of "S believes that Poirot had a moustache" which expresses a complete proposition and may be perfectly true. It is equivalent to: "S believes some ·Poirot had a moustache· computational."

To put the point a bit differently, harking back to section 6 above: Even though fictional etc. names lack referents, and even though (nonetheless) they are not semantically equivalent to descriptions, their internal counterparts in speaker-hearers of the language have functional roles that are *psychologically* analogous to those of descriptions. Thus we are able to cognize, internally, as if the names did abbreviate descriptions. Of course, this is also the case for denoting names; their internal counterparts function internally in ways parallel to the functioning of the empty representations, the semantic differences being entirely external. And according to Kinder, Gentler DR's two-scheme theory, attitude ascriptions bear interpretations that key on the computational roles rather than the semantic values of the representations they impute. That is why attitude ascriptions can be perfectly sensible and true even when their complements contain semantically meaningless empty names. And, I suggest, it is also why we hear fictional etc. sentences as meaningful.

I did insist, as even a Kinder, Gentler theorist must, that attitude ascriptions with empty names in their complements have interpretations that are "meaningless" or semantically defective. And I think that consequence of DR is correct. How *could* anyone believe *that Poirot had a moustache*--or "that Poirot [anything]"--if we are speaking of a propositional object in the truth-conditional sense? There is no such proposition. There is no such thing to believe.

It remains, as for Frege's Puzzle, to explain the dominance of the computational reading over the semantical reading for attitude ascriptions containing empty names. This time the explanation is evident. The semantical readings are semantically defective. Even an ounce of charity prevents us from hearing them.

That explanation would not serve for a speaker-hearer who is unaware that the names in question fail to denote. Suppose R, a friend of S's, ascribes to S the belief "that Poirot had a moustache," but like S, R thinks that Poirot was a real person. Then why do we evaluate R's utterance according to the computational rather than the semantical scheme? Well, *we* do so because we know that Poirot was not a real person, and we are being charitable ourselves. But why must R herself prefer the computational scheme in interpreting her own utterance, if her own concerns are not conspicuously functionalist or behavioral? She need not; she may prefer the semantical, and she may even be explicitly thinking of S's belief as *de re*. But she is mistaken. Since the name "Poirot" fails to denote and fails to make any other semantic contribution to the sentence she utters, it is, interpreted according to the truth-conditional individuative scheme, defective and not completely meaningful. R has underprivileged access to her own semantics.

Thus: When we know that an attitude complement contains an empty name, we automatically choose the computational scheme. When we do not know it, we *may* choose the semantical scheme, but if the complement does contain an empty name we therefore speak and hear defectively. When we are simply unsure, I suppose, we are more likely to choose the computational scheme depending on the strength of our suspicion.

I must not conclude before acknowledging a nastier consequence of Kinder, Gentler DR than has yet been revealed: Though (I have argued) the Harsh Doctrine can be made plausible in general, even I shrink from applying it to *negative existentials* in particular. It entails that "Poirot did not exist," "There was never any such English king as Arthur," and the like are "meaningless" or at least incomplete, sharing the semantics of open sentences like "$\sim(\exists x)(x = y)$." But surely, I concede, such sentences are true, and not merely things that can be believed when the computational individuative scheme is imposed.

The problem of negative existentials makes the Millian think more tolerantly of blank haecceities. That makes me, a Millian, think it is time to bring this chapter swiftly to a close. But there is a somewhat striking datum here, a meta-intuition: The Harsh Doctrine can be made fairly plausible, or so I have argued, for ordinary occurrences of empty names; but for the case of negative existentials in particular, it offends just as putridly as it ever did. That intuitive difference suggests that Frege, Russell and Sellars were right in holding "exists" to be a logically special verb, not a simple predicate of individuals, attaching to a higher-order expression rather than to a name.[36]

Unfortunately, candidates for the higher-order expression are not individually very promising. Three come immediately to mind: (i) A quote-name of our empty name itself. But the failings of the resulting metalinguistic approach are well known. (ii) A flaccid description. But we have rejected the bifurcation strategy. (iii) An abstract noun (e.g., "Holmesity") expressing a blank haecceity. I was right; it *is* time to bring this chapter to a close.

NOTES

[1] Marcus (1960, 1962), Kripke (1972/1980). See also Kaplan (1979) and elsewhere, and Donnellan (1970, 1976). Properly speaking, Kripke and Donnellan only attacked Description theories of names; as we shall see, it is a further inference to the correctness of any Millian theory, though many people from both sides of the issue have taken that inference for granted.
[2] There are a number of other arguments that I find unconvincing. I discussed them in Lycan (1980a); see also McKinsey (1978).
[3] Kripke (1979). Other criticisms of (I) may be found in Schiffer (1978), in Ackerman (1979b), and particularly Baker (1983).
[4] I argued this in Boër and Lycan (1975, sec. III), in Lycan (1978), and in Boër and Lycan (1980). See also Marcus (1961, 1962, 1981, 1983); Kaplan (1973, 1975, 1989); McKay (1981); Richard (1983); and the new "Direct Reference" theorists, Almog (1984, 1985), Salmon (1986, 1989), Soames (1987, 1988), Perry (1988), and Wettstein (1991).
[5] Devitt (1989) formulates a principle he calls "SP": "The meaning of a name is either descriptive or else it is [merely] the name's referent" (p. 207). He adds [p. 233, fn 19] that "Lycan['s]...'paradox of naming' starts with SP." That is a reasonable gloss on the opening paragraph of Lycan (1985), on which this chapter is based, but it is misleading in each of two ways: First, it suggests that I merely assumed SP, rather than defending it by appeal to the

Effability principle. Second, it is couched in terms of "meaning," a word I did not use; and that variation turns out to be important.

Devitt is concerned to attack SP, arguing as in Devitt (1981) for nondescriptive senses in the form of causal-historical chains. And he vehemently rejects Millian (or "'Fido'-Fido") theories as both implausible and, given his proffered *tertium quid*, gratuitous. But his claims are formulated in terms of "meaning":

Understanding a token requires assigning it the right meaning--in my terms, linking it to the same [causal] network that underlies its production. A person who gets this wrong will have misunderstood the token.... Yet the person may still have assigned the right referent. So assigning the right referent is insufficient for understanding. So having that referent is insufficient for meaning. (p. 229)

All this leaves open the relation between what Devitt calls "meaning" and *truth conditions*. The text (and conversation) make it clear that Devitt's causal chains are not being posited as parts of the truth conditions of unembedded sentences. Thus one might suspect that his dispute with the Millians is merely verbal, since as he acknowledges, his main opponents are concerned only with semantics in the narrow, truth-conditional sense.

However, he claims the causal chains do figure in the truth conditions of attitude ascriptions (p. 230; cf. Devitt [1981], Ch. 9), which accounts for substitution failure, and that certainly is a substantive disagreement at least with current "Direct Reference" theorists; see section 7 below. Unfortunately, I find the theory of Devitt (1981), considered as a strictly truth-conditional hypothesis, obscure, I think because it was not intended either to satisfy the Rule of Effability or to be considered as a strictly truth-conditional hypothesis.

[6] There is a further difficulty of how to fix the reference of "@" itself, pointed out by Austin (1979).

[7] This formulation is very closely related to Kripke's "Puzzle about Belief" (1979), though the case he concentrates on, involving his character Pierre, is more complicated. The solution I go on to propose would work for his case also, as well as (contra Devitt [1990]) his "Paderewski" example. It does not, or not obviously, work against Richard's (1990) excellent Phone Booth example; I sheepishly postpone discussion of that, for I agree it requires significant revision of the two-scheme theory.

[8] E.g., see Frege's footnote to p. 58 of (1966).

[9] I think this is generally agreed; however, Dummett (1973) contests it.

[10] Here is another way of putting the same point: Hintikka's semantical treatment of (A) requires that "Cicero" name one world-line and "Tully" name a diverging one. Therefore there is a nonactual world w at which Cicero and Tully are distinct. It follows from this by the definition of "rigid" that either "Cicero" or "Tully" is nonrigid in (A), and given our assumption that the two names have the same (type of) semantical status, they are both flaccid. In addition, corresponding to each of the two world-lines there is a function from worlds to sets: at each world the function spits out the unit set of the relevant individual manifestation. A function from worlds to sets determines a property, however complex or esoteric the property might be. Since our "Cicero" and "Tully" world-lines diverge, they determine distinct properties, at least one of which is lacked by our common referent at some world, which is to say that they determine different *contingent* properties of the referent. Thus, each name turns out to be semantically equivalent to the description whose matrix expresses the relevant contingent property.

(The foregoing is one natural reading of Hintikka's position. I should emphasize that his view could easily be modified in such a way as to avoid this consequence, though I think at the cost of collapsing it into Plantinga's option. Hintikka's world-line apparatus itself raises deeper skeptical questions about our naïve "rigid"/"flaccid" distinction, also; cf. Kraut (1987).)

[11] Two other theorists whose purported solutions to our dilemma collapse (I think) into already existing options are Ackerman (1979a, 1979b) and Stalnaker (1978, 1979a). Ackerman argues by way of a very demanding substitutivity test that although names undoubtedly have connotations, the connotations they do have are not shared by any descriptions. However, by her standards, many or most natural-language expressions have "connotations" that are not shared by any other expressions, and our interest shifts to her more permissive notion of the correct "analysis" of names; it turns out that for her names are equivalent to world-indexed descriptions that incorporate a causal-chain theory of reference. A name N, she says, is equivalent to (roughly) "the entity bearing R in @ to this very use of ⌜N⌝", where "R" denotes the appropriate sort of causal-historical relation. Thus, if I understand her correctly, she joins Plantinga in attributing a contradictory belief to our friend S.

Stalnaker proposes an ingenious appeal to two-dimensional modal logic, in an effort to determine what proposition is in fact believed by someone in S's position; Stalnaker is reluctant on the grounds of charity to charge S with self-contradiction. His independent motivation for his charitable interpretation of S's utterance is highly original (I discuss it in Lycan

[1980b]), but the interpretation itself is not--so far as I can see, he portrays *S* as ending up believing just the proposition that Russell would have said *S* believed. (However, Stalnaker's treatment applies only within intensional operators; he agrees with Kripke that names occurring outside such contexts are rigid.)

[12] I failed to appreciate this fully in Boër and Lycan (1980, sec. III), and spent only one page (451) on inadequate psychological handwaving.

[13] The somewhat unexpected distinction between (a) and (b) is enforced by Millikan (1984) and Dretske (1988).

[14] The best developed Causal-Historical theories I know are those of Devitt (1981) and Devitt and Sterelny (1987); see also Boër (1985).

[15] For purposes of this chapter I shall restrict discussion to "occurrent" or episodic beliefs, as *thoughts* are neologistically called. I am not happy with any account of "tacit" belief (see Chapter 3 of Lycan [1988]), and shall duck that issue here.

[16] On the teleological nature of functional characterization, see Lycan (1981, 1987, 1991).

[17] Perhaps the best example of this is Dennett (1978).

[18] I have in mind here Castañeda's (1966, 1967) data involving self-regarding attitudes, and the Twin-Earth cases adduced by Putnam (1975), Stich (1978) and Fodor (1980) in support of "methodological solipsism." I have discussed these in Lycan (1988, Ch. 4).

[19] These schemes are similar but distinct. For Sellars, linguistic roles (as marked by rules of assertibility, rules of inference, and the like) are essentially normative, while computational roles for Fodor are (I take it) more purely causal. My own preference is for a cautious fusion of the normative and the causal in the form of the teleological.

[20] What are "the representations associated respectively with" particular public names such as "Cicero" and "Tully"? Whatever inner states, I suppose, are typically activated by *S*'s hearing those names tokened and/or which typically figure in *S*'s production of those names.

Incidentally, the only explicitly Sellarsian treatment of proper names that I know of is Rosenberg (1978); he does not explore the two-scheme hypothesis.

[21] Field (1978, pp. 48, 51); Stich (1978); Fodor (1980, p. 67); Lewis (1981, pp. 288-289). Loar (1981, p. 117) makes much the same point I am pushing here.

[22] For extensive references, see Block (1986) and LePore and Loewer (1987).

KINDER, GENTLER DIRECT REFERENCE 169

[23] Incidentally, the two-scheme approach preserves Kripke's (1979) "disquotation" principle to the effect that S will sincerely and competently assert "P" only if S believes *that* P. For what that is worth; I have some independent doubts about the principle.
[24] A similar view of identity statements is put forward in Lockwood (1971). For further detail, see Boër and Lycan (1986), and section 8 below.
[25] Nor will I yet attempt to reply to a pair of important and telling objections made by Richard (1990), only one of which is his Phone Booth example aforementioned. For the record, I am inclined to think that the objections succeed and that my two-scheme view must be revamped, but I am not quite ready to undertake the task. I do not think the objections damage the defense of Millianism offered here.
[26] David Israel has pointed out in conversation that the term "Direct" is a misnomer in any case, for it suggests that the referential relation between a name and its bearer is *unmediated*, which is never strictly true; even the relation between an utterance of "that" and its referent is mediated by something. I suppose the idea was rather that reference is unmediated by anything *semantic*--in particular, names do not refer through semantical equivalence to descriptions--and the names' bearers figure in singular propositions themselves rather than by proxy.
[27] Murray Kiteley years ago called my attention to this feature of the Sellarsian semantics, and has emphasized the "motley" character of referential opacity in general (Kiteley [1981]).
[28] Indeed, Devitt has reminded me in discussion that they do not contribute *so much as* their own semantics, since their actual reference is irrelevant to the truth-value of the containing belief ascriptions. That is a disanalogy between belief ascriptions and sentence (G), since the name in (C) does introduce its bearer into (G)'s truth-condition. But the disanalogy seems harmless; it stems from the paratactic feature of the Sellarsian analysis, according to which belief complements are only exhibited and not asserted as part of the ascriptions that superficially contain them. (On the other hand, the disanalogy does take my theory farther from the letter of DR.)
[29] Of course someone could try to define a nonrelative sense of "informative" by binding the audience parameter with an existential or universal quantifier: "S is informative iff S would inform some audience of type T," or "S is informative iff S would inform any audience of type U." But I would not much want to join in the game of chisholming that ensued.

Yagisawa (1993) reminds us that Frege himself most definitely rejected every psychological interpretation of "informativeness"; by "informative" Frege meant roughly "a posteriori and not analytic." Yagisawa's own solution to Frege's Puzzle is based on that interpretation. I do not see the psychological and nonpsychological interpretations as mutual competitors, for identity statements can be informative in both senses, and each kind of informativeness needs explanation.

[30] Actually "merger" is a bit strong, suggesting total, final amalgamation and throwing away one of the folders. Devitt (1989) points out that even after one learns a new identity statement, one may continue to keep distinct files under the original two labels, that remain to some extent psychologically separate; think of "Hesperus"/"Phosphorus" and "Superman"/"Clark Kent." But in such cases the files do acquire a two-way connector, a conduit for substitution and conjunction inferences.

[31] Is the something *a proposition*? Not in the sense of the set of worlds that constitute the identity sentence's truth condition. There is some terminological slack concerning the relation between "that"-clauses computationally interpreted and propositions in the traditional sense; I shall not pursue the matter for now.

[32] Chiefly Salmon (1986).

[33] Salmon (1986, pp. 127-8) makes a few programmatic remarks. I find them congenial, and I daresay he would find my own view (to be defended a few paragraphs hence) reciprocally congenial.

[34] Moreover Braun (1992) presents some Kripke-style counterexamples to the claim that empty names are semantically equivalent to flaccid descriptions.

[35] More subtly, Parsons (1980) distinguishes between names of fictional entities and names that really are empty in that they fail to refer even to Meinongian Objects.

[36] For a defense(!) of Russell's peculiar version of this thesis, see Lycan (1994).

CHAPTER 8

RELATIVE MODALITY

We have had a good deal to say about the nature and role of possible worlds themselves. But we have barely mentioned that further element of any Kripke model, the accessibility relation. The accessibility relation is important to modal logicians because the formal features ascribed to it by a particular semantic interpretation explain the truth of the characteristic axioms of the formal system being interpreted. But my purpose in this chapter is to show that variations in accessibility are, if anything, even more important linguistically, and philosophically. In particular, I shall give several examples of knotty philosophical issues that become more tractable once they are approached as questions about accessibility.

No ordinary English sentence expresses an unrestricted alethic modality. Rarely do we hear mention of logical necessity, logical possibility or entailment outside a philosophy department. Indeed, many undergraduates have trouble grasping the concept of barely logical possibility, even with professional help.

Rather, all everyday modalities are restricted, relative to contextually determined sets of background assumptions.[1] That is both unsurprising and a very good thing. Interestingly, though, few of even those street-level restriction classes themselves correspond to recognizable philosophical categories.[2] As a case very much in point, the concept of *natural or physical* necessity is hardly better known to everyday English than are the purely alethic modalities. I shall give some examples of real restriction classes.

CHAPTER EIGHT

1. MODALITIES IN NATURAL LANGUAGE

The syntactic behavior of what have traditionally been called "modal auxiliaries" in English reflects the drastic semantic/pragmatic context-dependence of modal notions.[3] Sometimes it is revealed in an unexpectedly specific though still tacit restriction: "That woman speaks eighteen languages and can't say 'No' in any of them" (attributed to Dorothy Parker). Sometimes it is revealed in an oxymoron: "I have to, but I can't."[4] For some further examples, consider the following passage from a rather good trash novel of the 1960s:[5]

Boo and Ira have been conducting a parodically torrid adulterous affair. They are now talking by 'phone.

"I have to see you, Ira. As soon as possible. I'm going out to the beach house this evening at six o'clock to talk to a man who's building a new jetty. I'll surely finish with him in a half-hour or so. Could you meet me out there at seven o'clock?"

"How can I, Boo? At eight o'clock I'm doing the telecast from Acanthus."

"That's right, I forgot," she said. "All right, then, how about this: I'll wait at the beach house and you can meet me there when you finish the telecast. Okay?"

"But you'll miss the program. You don't have a tv set at the beach."

"I know, Ira, and I'm terribly sorry. But this is more important. Please, *please*, my darling, I *have* to see you tonight!"

Elation came galloping back to Ira. "Yes, yes, tonight!" he cried. "Tonight! Yes!"

"At the beach house."

"Yes, my darling. Yes, my good, sweet-- " Elation made another quick retreat. "What do you mean, you 'have to' see me?" he said warily. "Do you mean it's like an *emotional* need? For example, when two lovers are kept apart for a length of time, there is a certain psychic compulsion to be reunited. Is that what you had in mind?"

"Ira, I can't talk any more. There are people in the house."

"Wait, wait, wait!" he insisted. "I just want to clarify this one point. It boils down to those words 'have to.' Could you possibly be a little less ambiguous?"

At this point Boo breaks off the conversation and hangs up. She had earlier decided to break off the affair; but five chapters later, the reader is not surprised to learn that Boo now "had to" see Ira to tell him he had made her pregnant.

Now consider just the *squarely* modal locutions in the foregoing passage (underlined and indexed):

"I have to$_{[1]}$ see you, Ira. As soon as possible$_{[2]}$. I'm going out to the beach house this evening at six o'clock to talk to a man who's building a new jetty. I'll surely finish with him in a half-hour or so. Could$_{[3]}$ you meet me out there at seven o'clock?"

"How can$_{[4]}$ I, Boo? At eight o'clock I'm doing the telecast from Acanthus."

"That's right, I forgot," she said. "All right, then, how about this: I'll wait at the beach house and you can$_{[5]}$ meet me there when you finish the telecast. Okay?"

"But you'll miss the program. You don't have a tv set at the beach."

"I know, Ira, and I'm terribly sorry. But this is more important. Please, *please*, my darling, I have to$_{[=1]}$ see you tonight!"

Elation came galloping back to Ira. "Yes, yes, tonight!" he cried. "Tonight! Yes!"

"At the beach house."

"Yes, my darling. Yes, my good, sweet-- " Elation made another quick retreat. "What do you mean, you 'have to' see me?" he said warily. "Do you mean it's like an *emotional* need$_{[6]}$? For example, when two lovers are kept apart for a length of time, there is a certain psychic compulsion$_{[7]}$ to be reunited. Is that what you had in mind?"

"Ira, I can't$_{[8]}$ talk any more. There are people in the house."

"Wait, wait, wait!" he insisted. "I just want to clarify this one point. It boils down to those words 'have to.' Could you possibly$_{[9]}$ be a little less ambiguous?"

Here we have a wondrous profusion of distinct modalities, every one alethic. (I shall not more than mention the additional "surely," "how about...?," "important," "Okay," "kept," or "boils down to," much less the extraneous propositional-attitude locutions and indirect discourse.) To catalogue and analyze just the nine squarely alethic and plainly distinct modalities found in the passage would require at least a book-length effort;[6] I shall only gloss them here. (The glosses will include further alethic terms, with no end in sight.)

(1) "I have to see you...." My seeing you is required by some contextually specified external circumstance (not just the pregnancy, but Boo's reason(s) for informing Ira of the pregnancy); the modality itself is (further) relativized to whatever psychological and/or social norms apply. A very long story, though in part because drastically underdescribed.

(2) "...as soon as possible." As soon as is consistent with...? The modality here centers on Ira and both the norms and the circumstances centered on him; but I think it is also constrained by Boo's situation and norms applying to that.

(3) "Could you meet me...?" Possibly the same as (2), but I doubt that. "Possible" in (2) grammatically applies to the infinitival complement of "I have to see you," or some suitably nominalized version of it such as "my seeing you"; but "Could" in (3) is the head of a VP predicated directly of the subject "you" (= Ira), suggesting a closer connection with Ira's own circumstances and less to do with Boo's situation.

(4) "How can I, Boo?" Someone might think that (4) is the same modality as (3), but the identity is unlikely. "Could" in (3) and "can" in (4) are not related as past tense and present. Rather, "could" is subjunctive where "can" is indicative; and (as is conceded by all parties, whatever may be the deeper analysis) the subjunctive introduces a wider range or larger space of possibilities than does the more actuality-oriented indicative.[7] Perhaps in a more remote sense Ira *could* meet Boo, but as things are he *cannot*. Yet this "can" has nothing to do with Ira's *abilities* or even with his legal situation; it hovers somewhere between his institutional responsibilities (to the television network that employs him) and simple prudence. N.b., the negative answer to Ira's rhetorical question is something that Boo is mutually presumed to know but admittedly has forgotten, and it trumps her more remote "could"; they are neither disagreeing nor talking past each other.

(5) "...[Y]ou can meet me there...." In the absence of further context, this is a tough one, but I hear it as well paraphrased by something between the same "can" as (4) and "[Then(?)] it's OK [with me] if you just meet me...." It is not *simply* the same "can" as (4)'s, which occurs within the scope of "How about...?"

(6) "...an *emotional* need [to see me]." The concept of "need" is tricky, but I think White (1975) is correct in maintaining that any felicitous token of "need" tacitly refers to a contextually indicated goal or end-state, and the idea is that the needed state of affairs is necessary for the realization of that end-state. (There is no particular universal or even default end-state.[8]) "Emotional" here begins to classify the relevant end-state envisaged by Ira. But the underlying sort of "necessity" is vexed and (at least) needs[9] much further unpacking.

(7) "...a certain psychic compulsion...." Here the modality occurs as a noun, but the noun is a derived nominal (from "X compels Y") and needs[10] an underlying subject, an agent or other being that does the compelling—here, presumably, Boo's psyche conceived as necessitating her reunion with Ira by *acting on* her.[11]

(8) "...I can't talk any more." This, suddenly, has nothing to do with Ira's situation and proceeds entirely from Boo's immediate circumstances.[12] Her talking any more would be (as things are) incompatible with privacy, or with discretion on her part, or with prudence. (Those are the candidates nominated by the novel's surrounding plot. But others are at least consistent with the portrayed facts: She may be flat out of emotional energy; her sanity, or his, might be at stake; an armed guard might be about to grab the phone; etc.)

(9) "Could you possibly be...?" Another toughie, in part because we are not sure whether Ira is being ironic. If we assume he is not, we still need to know whether "Could...possibly" is (a) a merely emphatic redundancy, (b) a compounding of two distinct modalities, or (c) something in between, such as a widening of the range of the first modal operator by the second. In any case Boo's presumed constraints are quite indefinite. They are unlikely to be the same that control the modality in (9); a direct reply to (8) would refer more directly to the presumed constraints, rather than to the semantic

interpretation of her utterance (as, e.g., "Could you possibly hang tough and also put a few more quarters in the 'phone?").

Again, though the foregoing comments are intended to be roughly true, they are at best impressionistic; not one is put forward as having much analytical value.[13] My purpose is only to indicate that everyday English is shot through with restricted alethic modalities whose restrictions are almost capriciously diverse, rarely aligned with any easily specifiable modal concept known to logicians, and irreparably vague--yet calculated on the spot by ordinary human speakers/hearers with hardly a conscious thought. The rate of modal montage exhibited in the Boo-Ira exchange is remarkable, and the linguistic competence involved in our rapid interpretive shifts is simply amazing.

Notice particularly that the philosophers' idea of nomic or physical necessity plays no role here. Genuine laws of nature are thin on the ground in any case,[14] and surely none of them applies to matters such as Boo's reasons for informing Ira of the pregnancy, Ira's institutional responsibilities, Boo's eventual reunion with Ira, her privacy and discretion, etc. (even if those items are supervenient causes or effects, connected under lower-level descriptions by the laws of some natural science). Rather, the modalities in question seem representable only by accessibility relations that are themselves unphilosophical and vague. I suspect that that vagueness is part of the human condition.[15]

Further and often even more idiosyncratic examples are provided by the "-ble" suffix attaching to verbs. Consider: (1) A person commits an *unspeakable* act (howler, solecism, *faux pas*, outrage,...). That means the tale cannot be told without--I would say-- scandal, or the mention of things whose very mention offends propriety; it certainly does not mean that no one has the ability to tell it, or even that telling it is illegal.[16] (2) A person says something *unprintable* or *unrepeatable*; that normally means the person used taboo words, though "unprintable" might alternatively mean something forbidden by journalistic ethics. (3) Physical things and stuffs can be *inedible* in various ways--but usually not in the sense that such a thing simply *cannot* be eaten, as holds for the Eiffel Tower or the Earth itself. Cyanide is inedible in that one cannot eat it *and survive*. A used milk carton is inedible in that eating it has no nutritional value and may cause one some considerable intestinal distress. Mice are classified as inedible creatures in that, even though they are nutritious, eating one is (I am not sure which:) potentially unhealthful, or disgusting, or socially proscribed, or perhaps otherwise a *very bad idea*. A local greasy spoon's food is inedible in that one cannot normally even make oneself eat it, though one might manage to gag it down at gunpoint; but a pretentious restaurant's food might be inedible only in that its low quality is grossly incompatible with the restaurant's pretensions and prices, and/or with the gourmet's minimal standards. (4) Pharmaceutical companies advertise "chewable" aspirin. In context, this could hardly mean "can be chewed," or even "can be chewed without such-and-such physical damage"; it means something like, "can be chewed without seriously offending the taste buds."

This list could go on and on and on. I trust that, once the reader has got the idea, he/she will see the phenomenon everywhere. Indeed, a careful reading of this very chapter will yield scads of examples, and not because I tried to put them there.

Notice further that some alethic modalities are not reflexive. (It is well known that deontic and doxastic modalities are not reflexive, since what ought to obtain and what is believed to obtain may not in fact obtain, but one would naturally think that any *alethic* modality would support $\Box p \vdash p$ and its dual, $p \vdash \Diamond p$.) The previous example, "I have to, but I can't" already demonstrates the point, but here is a striking example from a biography of Henry Kissinger:[17]

Later, Kissinger would wonder about the interview [that he had given to Oriana Fallaci on November 4, 1972] and sigh, "I couldn't have said those things, it's impossible." But he never denied it.

Finally, do not forsake this book without noticing some purely alethic modal concepts that pervade everyday discourse but are virtually untouched by logicians and philosophers of modality. A blazing example is the concept of *luck*, including the notion of an

178   CHAPTER EIGHT

event's or a thing's or a person's being lucky. I know of just one paper dedicated to this topic, viz., Rescher (1990); but there is a fat and big-selling book[18] waiting in Plato's Heaven, to be plucked and/or published by any competent philosopher who can spare the time. The same is true of the concept of *accident* (of which more shortly). Then there are "coincidence," "happenstance," "chance," "alternative," "scenario," "ensure," "guarantee," "compel," "force," "prevent," "forestall," "allow," "permit," "afford," and countless other alethic expressions, no two the same (n.b.). I wish I were younger and had a bigger grant.

2. IMMEDIATE PHILOSOPHICAL RETURNS

Once the lunatic relativity of modal terms is recognized, we are able to turn a clearer eye onto a number of surprisingly disparate traditional philosophical issues. When we do so, those issues are shown to dissolve or at least to become more tractable.[19] I shall catalogue merely a few.

(i) *Possibility "in principle."* Philosophers are notorious, and often hated by laypersons, for their appeals to what is possible "in principle." One of our own, Hilary Putnam (1978, pp. 63-4), has paused to take us professionally to task:

'Is it possible "in principle" to do X?' typically means 'Is X physically possible (or, perhaps, logically possible) to do?' But this is not what 'possible in principle' *should* mean in discussions of topics such as... [, e.g., the precise prediction of human behavior].

Putnam goes on to argue that philosophers' claims of possibility-in-principle, where only logical or nomological possibility is meant, are seldom interesting or useful.

But once we appreciate the fickleness of the accessibility relation, we see that the notion of possibility-in-principle can always in principle be made fairly precise *and* useful. By its nature, it is possibility of an avowedly inclusive grade, that outruns present circumstances and abilities. But how far it outruns them depends on contextual purposes. Conservatively, it could mean "scientifically possible though beyond our present or expected equipment and grants." Or it could mean "possible *in real time*," a concept of concern to Putnam. More loosely, it could mean "theoretically possible [modulo the relevant science] though beyond any presently imaginable device to bring about." It could mean "theoretically possible though only because of quantum indeterminacy" (as that my Compaq should spontaneously leap to the ceiling and then fly figure-eights around the room). It could mean "conceptually and mathematically possible though in violation of a law of nature."

For what they are worth (almost nothing), my own linguistic and modal intuitions seem to pull up at this point. It would be hard for me to call a *mathematical* impossibility "possible in principle." But if one's purposes were so abstract as to ignore mathematics--perhaps they are "purely conceptual" in analytic philosophy's intended sense of that term[20]--then contramathematical fictions might be counted as possibilities "in principle." I am less sure about contra-"conceptual" possibilities such as round squares and male vixens.

In any case, the point about "possibility in principle" is that that phrase is perfectly serviceable, so long as context makes clear *what* principle is in question.

(ii) *The notion of "accident."* That notion has figured in philosophy as well as in everyday life. A paradigm case is its role in epistemology's postGettier or "JTB+" literature, whose project is to find some "fourth condition" of knowing and thereby block Gettier's (1963) counterexamples to the traditional "justified true belief" account. Though many Causal and Reliability theorists of knowing have been guided by the same intuition, the paradigm case of an "accident"-based theory of knowing is Unger (1968).

Unger proposes what might unpejoratively be called the "Naïve approach" to the Gettier problem. Upon first hearing a Gettier case described, students commonly respond by saying it is "just an accident" that the Gettier victim's belief is true. Thus Unger:

> For any sentential value of *p*, (at a time *t*) a man knows that *p* if and only if (at *t*) it is not at all accidental that the man is right about its being the case that *p*. (p. 158)

Unger carefully distinguishes the man's *being right* from the man's *coming to know* (p. 159), and, more interestingly for our purposes, "not at all accidental" from the weaker "not an accident" (pp. 161-62)

I believe there is much to be said for the "Naïve" idea, and it has never been properly assessed, that is, assessed on its own terms. (Both Causal theorists and Reliabilists have claimed Unger as an early member of their respective camps, but he was not.) The tool most urgently needed for the task is at least a schematic analysis of the notion of accidentalness.

It seems clear enough that "accidental" expresses an alethic modality, and a relative one at that. For the sake of discussion, then, let us accept the appropriate schema (harmlessly conflating use and mention): "It is accidental that P" will be explicated as "Γ does not entail P" (i.e., at some world, Γ obtains but P does not), where "Γ" designates some contextually determined body of lore or presumed fact.

But on this analysis, a penetration principle is demonstrable: If it is accidental that P and Q entails P, then it is accidental that Q. (If there exists a Γ-world *w* that is not a P-world but every Q-world is a P-world, then *w* is not a Q-world either.) Any apparent counterexample to the principle must result from a shift in the value of the parameter Γ between premise and conclusion. Of course, Γ does shift its value easily; the notion of accident is highly perspectival.[21]

Unger almost explicitly rejects the penetration principle for accidentalness; he insists (p. 160) that it can be not at all accidental that a subject is right about something even though it is quite accidental that the subject even exists. And, much more to the point, the key examples against which Unger tests his analysis are apparent violations of the principle. E.g.:

> [A] man may know about an auto accident: when the car accidentally crashes into the truck, a bystander who observes what is going on may well know that

the car crashed into the truck and accidentally did so. He will know just in case it is not at all accidental that he is right about its being the case that the car crashed into the truck and accidentally did so. (p. 159)

Here the parameter shift is obvious: when Unger says it is not at all accidental that the man is right about the auto crash, he means it is not at all accidental *given that the crash did occur*. Thus, the fact that the crash occurred is (obviously) not a background condition relative to which the crash itself was accidental, but is a background condition in the set relative to which the man's being right is nonaccidental. Γ takes two different sets of conditions as values here--parameter shift *par excellence*.

Similarly (p. 159), it may be quite accidental that "a man" have the belief that he will be fired, say, because he chanced to overhear his employer instructing the payroll office. By the penetration principle it would follow that his being right is accidental, since his being right entails his having the belief; but Unger is at pains to deny that, since he judges that the man knows he will be fired. I daresay Unger means that it is not accidental given that the man does have the belief; *that* background condition is what shifts the parameter in this case.

By now the reader will have noticed that the penetration principle both is valid and is not valid, even on my own understanding of "accidental": It is valid in that *if the parameter Γ is held fixed*, there cannot be true premises and false conclusion. It is invalid in that there are triples of English sentences, all perfectly correct to utter and true in the same context, that constitute true premises and false conclusion, though by dint of tacit parameter shift. Unger is quite right to reject the penetration principle in this second sense.

But in light of that, he will have trouble with a further case, his own anticipated worst apparent counterexample (pp. 167-70): "A man" dreams that Schimmelpenninck will win the 1965 Kentucky Derby; the man's veterinarian friend hears him mutter this prediction upon awakening, an infallible sign that that is what the man has dreamt and that he irrationally believes it. The altruistic veterinarian then very reliably sees to it that Schimmelpenninck does win, by sedating all the other horses entered in the race. Unger argues that, contrary to appearances, it is still somewhat accidental that the dreamer is right about Schimmelpenninck winning, for at least two reasons: First, "...why did the veterinarian make just that particular resolve?" Second, the dreamer's belief only happens to have been one that was within the veterinarian's power to make true.

But a treatment parallel to the ones Unger himself seems to have intended for the auto crash and employee cases would produce the opposite result: Given that the dreamer does believe the horse will win and given that the veterinarian has in fact carried out his resolve, it is not at all accidental that the dreamer is right about the horse winning. Yet the dreamer does not know; still a problem for Unger's analysis.

The bad news is that on our crude schematic analysis of accidentalness, Unger's discussion does not get off the ground unless parameter shift is allowed; and even if we try to accommodate parameter shift, lots of work would have to be done to chart Γ's behavior from context to context, in order for us to tell even roughly what predictions Unger's theory makes and what ones it does not. If we are to get anywhere in evaluating the "Naïve" approach to the Gettier problem, we should start with an independently established ethology of Γ and only then apply the resulting theory to Unger's formula. The good news is that the latter project is viable, indeed an excellent research topic.

The "Naïve" response is still intuitive but by no means simple; we are off and running.

(iii) *The concept of "miracle."* Swinburne (1970) begins with the sentence, "There are many different senses of the English word 'miracle'..." (p. 1). Like Hume, Swinburne starts his investigation with the notion of *a violation of a law of nature* [deliberately brought about] *by a god*" (p. 11). The problem is obvious even before we get to "brought about" or "a god": Reputedly, a *law of nature* is at least a true universal generalization with some modal force attached; the generalization holds not only at @ but at every suitably "nearby" world. Yet if a *violation* actually occurs, then (by the predicate calculus and even by syllogistic) the relevant generalization is falsified. So "violation of a law of nature" is on its face a contradiction in terms.

Two basic moves are available to those philosophers who wish to preserve the concept of miracle as violation of law.[22] We can weaken the concept of *violation*: One might insist that the real natural laws of our universe are exceptionless, and so not violated in the strict sense, but are so only in virtue of containing surds or kinks that accommodate the particular miraculous events that have occurred or ever will occur. A law unimpeded by any actual miracle will be neat and fully general, just as laws are conservatively taken to be, but a putative law that is falsified by a miracle, as was the conservation of angular momentum by Joshua's having made the sun stand still in the sky, is inaccurate; and the correct version must contain the appropriate specific, single-exception clause ("Angular momentum is conserved, except on [date and time], before Jericho, when Joshua stopped the sun..."). Such single-exception clauses offend our intuitive concept of law, certainly, but we are talking of miracles here.

Lewis (1979, pp. 468-69) makes the move less unattractive by suggesting a positive account of miracles that preserves some integrity for the "law" that gets violated:

...the violated laws are not laws of the same world where they are violated.... I am using 'miracle' to express a relation between different worlds. A miracle at w_1, relative to w_0, is a violation at w_1 of the laws of w_0, which are at best the almost-laws of w_1. The laws of w_1 itself, if such there be, do not enter into it.

Thus Lewis rejects kinky laws, but weakens the notion of "violation" in his own corresponding way, by preserving the idea that the laws of the home world w_1 are not themselves violated in the sense of being falsified.

Since every event that occurs at any world violates the laws of *some* other world, considerably more needs to be said about the relation between w_1 and w_0; the key notion in Lewis' passage is that of an "almost-law." What is it for the laws of w_0 to be the almost-laws of w_1? Presumably, on Lewis' famous Similarity theory of distance between worlds, the relation involves similarity in some respect, however complex.[23]

The other possible basic move is to weaken the concept of *law*: Following Smart (1964, Ch. 2), Swinburne (p. 27) removes the requirement of *truth*, and weakens the "law" concept to admit "unrepeatable" exceptions. We could ask whether that "-able" is metaphysical (surely not, if the miracle has already happened once) or epistemic (possibly, since as Swinburne points out, miracles may slightly weaken the predictive power of the generalizations we thought were strict laws, but *leave us with nothing better*, so that we have no choice but to go on trusting them). But Swinburne removes the doubt by defining[24] a "repeatable" exception as one which "would be repeated in similar circumstances," by which he evidently means *naturally* rather than supernaturally similar circumstances; the epistemic consideration mentioned in my preceding sentence is only an answer to the question of how we might *tell* whether an exception is "repeatable."

But either way, the miracle forces allusion to laws that do not hold at @--for Lewis, because they are laws of an only nearby world, and for Smart and Swinburne, because although they are laws of @, "laws" need not be true. We could join Lewis in focussing on truth, and insist that the violated laws be those of a "nearby" world (bearing with the vagueness of that "nearness" or "similarity" relation), or we could focus on a strong notion of lawlikeness and insist that the "laws" of @ that are falsified are perfectly *lawlike* even though they are false at @ (leaving the vagueness in our concept of "law," once universality has been abandoned).

I am not entirely sure that in the end the difference is more than notational. Each of the two moves presupposes the same picture, that of a world neatly axiomatized by a set of basic physical principles marred by some fairly rare and isolated exceptions. Lewis chooses to reserve the term "law" for application to kinkless generalizations, weakens the notion of violation, and tacitly appeals to a similarity relation that ignores the kinks; Smart and Swinburne willingly apply the term to kinky falsehoods but invoke the further modal concept of "unrepeatability." Kinklessness seems to be the core idea; putatively actual kinks are accommodated by Lewis' metaphysical similarity relation or by Swinburne's counterfactual "would [not] be." But on Lewis' theory of counterfactuals, the latter mobilizes a similarity relation of its own. Despite the problems I mentioned in fn 23, the two similarity relations may prove to be one and the same, and the two moves only terminologically distinct.[25]

Whether or not the moves differ substantively, each can be characterized in terms of the accessibility relation. Smart and Swinburne allow "laws" to be false at the worlds of which they are laws. If laws are what is nomically necessary, then a law of *w* must be true at every world that is nomically accessible from *w*; but that means nomic accessibility is nonreflexive. By contrast, Lewis' move preserves reflexivity, for on his view a law must still be true at its containing world. This seems to me to be an advantage, for as I remarked in sec. I, it is natural to think that any *alethic* modality, unlike doxastic and deontic modalities, would be reflexive.

True, as I insisted in the same breath, there are ostensibly alethic modalities that we already know to be nonreflexive. But the examples I gave ("I have to, but I can't" and Kissinger's "I couldn't have... [but I tacitly concede that I did]") presumably involve normative and personal factors that one does not associate with anything so general as a *law of nature*. I shall not try to make this point any more conclusively, for since the two moves differ so little in substance, only a small advantage is needed.

3. MORAL PSYCHOLOGY

I shall mention just three further issues, each important in moral psychology: The compatibility of human freedom with causal determinism; "'ought' implies 'can'"; and the ancient question of "why we should be moral."

(iv) *Free will and causal determination.* Here almost all my work has already been done for me, by Slote (1982). Slote expounded and criticized what is probably the most powerful existing argument against freedom-determination compatibilism. Versions of the

argument have been given by Ginet (1966), Wiggins (1973), van Inwagen (1975, 1983), and Lamb (1977); Slote dubs it "GLVW."
GLVW takes the form

$$\Box S$$
$$\Box(S \supset A)$$

$$\therefore \Box A,$$

where A is the performance of an arbitrarily selected future action of mine, S is a total efficient cause of A existing before I was born (here we assume determinism for the sake of argument), and the box is interpreted as "unalterability" of some sort. The premises then seem entirely plausible. And the form is valid for any standard alethic modality; its corresponding conditional is what Slote calls the "main modal principle" (MMP). He traces the appeal of this principle to two presumed formal features of whatever box is intended: "agglomerativity," or closure under conjunction introduction, and "closure" proper, viz., under logical implication.[26]

Slote argues that the MMP fails for some modalities, and that, surprisingly, it fails for some squarely alethic modalities in particular.

> ...some forms of relational [= relative] necessity can arise only in certain narrowly circumscribed ways; and when restrictions on the way a given kind of relational necessity can come into being unhinge it from agglomerativity (or closure or... [the MMP]), we may say that such necessity is *selective*. (p. 13)

Just as agglomeration and perhaps closure fail for obligation in a rather narrow sense that Slote discusses (pp. 11-14), they fail for *nonaccidentalness* in the colloquial sense. Nonaccidentalness "comes into being" through a "routine plan," or as we might say, some sort of "coherent background" (CB) that is local in scope--a rational *story behind* the nonaccidental occurrence. Agglomerativity fails because there may be a CB behind e_1 and another CB behind e_2, but no third

CB behind the composite occurrence $e_1 + e_2$, because the two existing CBs do not amalgamate to form a suitably coherent rational story; thus $e_1 + e_2$ may be accidental even though neither e_1 nor e_2 is accidental.

Slote now suggests that GLVW's notion of "unalterability" is a similarly selective necessity, i.e., a necessity that comes into being in a special way that "unhinges it from agglomerativity." What is that special way, and what is the particular kind of factor that gets "selected," the analogue of a CB?

According to Slote the factor is large and diffuse. If I understand him correctly, he thinks the unalterability-by-me of circumstances obtaining before my birth is a matter of their having a total efficient cause that is itself unaffected by my beliefs, desires, intentions and other conative structure (hereafter just "my CS").[27] So as the "selected factor" he nominates the total efficient cause itself, or perhaps the-total-efficient-cause-as-unaffected-by-my-CS. Laws of nature are also unaffected by my CS, though for a different reason.

But now we can see why the MMP fails for unalterability-by-me. GLVW's two premises are true, for the reasons just mentioned. But the conclusion "□A" does not follow, for since A is a future action of mine and I am now part of A's total efficient cause, the total efficient cause is *not* unaffected by my CS; in fact, my CS is a rather prominent part of it. To be unalterable-by-me is to be determined or necessitated *in a certain way*; the past state of affairs S and the laws of nature are necessitated in that way, but my future actions, though they are necessitated, are not necessitated in that way. For the same reason, either agglomerativity fails or logical closure does (p. 21).[28]

This all seems right to me. But it needs some further explication. Let us return to the example of accidentalness. Slote (p. 15) offers an example of an accidental meeting in a bank between Jules and Jim. Each party has been very carefully sent by his superior to the bank, "as part of a well-known routine or plan of office functioning"; thus, it is no accident that Jules is in the bank at the time in question, and no accident that Jim is there then. But it is entirely accidental that both are there at the same time.

So: Jules' presence in the bank was necessitated by CB1 and Jim's presence there was necessitated by CB2, but there is no CB that necessitated Jules'-presence-and-Jim's-presence, because in particular the amalgam CB1+CB2 is not itself a CB.

$$(w_{CB1})In(w, Bs)$$
$$(w_{CB2})In(w, Bm)$$
$$\overline{}$$

-- but nothing of the form $(w_{CBn})In(w, Bs \& Bm)$ follows, because there is no such CBn. (Perhaps more accurately, instead of portraying the sentences as tacitly referring to particular CBs, we should construe them existentially:

$$(\exists CB1)(In(x, CB1) \supset In(w, Bs))$$
$$(\exists CB2)(In(x, CB2) \supset In(w, Bm))$$
$$\overline{}$$

but nothing of the form $(\exists CBn)(In(x, CBn) \supset In(w, Bs \& Bm))$ follows, as before.)

The Slotean exegesis of van Inwagen's argument is parallel. To be unalterable-by-me" is to be necessitated in a certain way, viz., by a particular factor." The factor is determining, and determining in a way that simply bypasses or circumvents my conative state CS; call such a factor a me-excluding determinant" (MED). But the MEDs differ as between the first premise, $\Box B$, and the sencond, $\Box(B \supset A)$. The prior state B is necessitated by, as we might say, its own past, the latter excluding my CS because I neglected to exist at the time. The conditional truth $B \supset A$ is necessitated by the laws of nature, which exclude my CS because $B \supset A$ follows deductively from the laws alone regardless of anything about me. Thus we have

$$(w_{B's\ PAST})In(w, B)$$
$$(w_{LAWS})In(w, B \supset A)$$
$$\overline{}$$

--but nothing of the form $(w_{MEDn})In(w, A)$ follows, because there is no such MEDn. In particular, the amalgam B'sPAST+LAWS is not a MED, for it does not determine the action A in a way that bypasses my CS; though it does determine A, it does so only by passing through, and by means of, my CS. In general, even where Z is entailed by the conjunction X & Y, that X is determined by a MED and Y is determined by a MED does not entail that Z is determined by a MED. So □A cannot be derived. (As with nonaccidentalness, we might better reconstruct the argument using existential quantification, rather than positing tacit reference to the particular MEDs, but the result would be the same.)[29]

To make a more general point, note that incompatibilist arguments typically try to take any alethic sort of necessitation and make it into *compulsion*[30]--a powerful rhetorical strategy, since we all know compelled acts are not free. (Actually we know nothing of the sort. In the loosest sense of "compel" in English, I can be compelled by people or by circumstances to do things which I nonetheless, at the same time, freely choose to do, as when I hand over money at gunpoint, give in to blackmail, or merely follow orders.) Speaking in terms of relative modalities, the reason compulsion does not follow from bare necessitation is that as I noted in sec. I, the notion of compulsion introduces an actual agent or quasi-agent who does the compelling; though I may buy a Hohner F at every physically possible world satisfying some further general condition C, and even in every such world whose temporal past approximates that of @, there need be no actual agent or quasi-agent N and action A of N's such that at every physically possible world at which N does A, N's doing A causes me to buy a Hohner F.

(v) *"'Ought' implies 'can'."* Almost everyone has accepted some version of the doctrine, best put contrapositively:, that if it is simply not in my power to do such-and-such, then I am not in any strong sense obligated to do that thing (it is conceded that there may be some ossified or technical sense in which I am so "obligated"). Indeed, many philosophers have taken the slogan, "'Ought' implies 'can'," to express an obvious truth, whichever truth that might be exactly.[31]

At the same time, however, the literature teems with well-known examples whose protagonists are unable to do what they ought to do. Certainly the casual sentence "I ought to, but I can't" is often perfectly acceptable in context--even when the grade of "impossibility" being invoked falls far short of anything like nomic impossibility or even powerlessness in the circumstances.

As White (1975, Ch. Ten) maintains, we face a task of mediation here, that of explaining the philosophical plausibility of "'Ought' implies 'can'" while preserving an adequate exegesis of the apparent counterexamples. Obviously, this could be done either (a) by rejecting "'Ought' implies 'can'" and exhibiting its plausibility as specious, (b) by holding into "'Ought' implies 'can'" and explaining away the counterexamples as merely apparent, or (c) by finding an interpretation of "'Ought' implies 'can'" under which it remains true even though the examples are indeed cases in which, on a distinct but equally viable interpretation, someone cannot do as he/she ought. Before trying to choose between these, let us survey some of the apparent counterexamples.

(i) It is painfully obvious, and only to be gotten out of the way, that cosmic "ought-to-*be*"'s do not imply "can"'s predicated of actual individuals. "There ought to be a permanent end to war of any sort" hardly implies that some individual is able to put an end to war; it does not imply even that the human race taken collectively has that ability. The case of "There ought to be no horribly painful diseases" is even clearer. All that such "ought-to-be"'s mean is that no perfect (= deontically accessible) world contains wars after the date in question, or painful diseases, or whatever.

(ii) Even some "ought" sentences with animate subjects have the ring of "ought-to-be"'s rather than "ought-to-do"'s; it is clear that they do not imply "can"'s predicated of the subjects. "George ought to be living in the twenty-fifth century"; "Hitler ought never to have been born" (Ziff [1987]).

Cases of types (i) and (ii) can easily be written off by a defender of "'Ought' implies 'can'," for it is unlikely that the doctrine was ever intended to apply to "ought-to-be"'s; rather, it means "'Ought-to-*do*' implies 'can do'." But:

(iii) Christians believe of certain character flaws that we ought not to have those flaws and ought to take steps to rid ourselves of them, but they also admit that the same flaws are built into human nature, in so basic a way that even Christ Himself found it psychologically impossible to avoid manifesting a few of them. Thus, on this view, a morally perfect human character is doomed to remain an unfulfilled regulative ideal. (Indeed, I suspect almost anyone's convictions about character have that same consequence.) To some optimists, any such view may seem implausible, repugnant, unfair, or all three, but it is not *incoherent*, at least not obviously so.

(iv) Similarly, we sometimes ought to rid ourselves of undesirable personality traits. I ought to become less compulsive, more domestic, neater, and (I have been told[32]) less vain. But it may well be outside my power to change myself in all or even any of those ways.

(v) We ought not to have certain emotions or feelings; e.g., we ought not to feel resentment or personal blame against people who we know have not wronged us in any way, and we ought not to be glad at someone else's misfortune (unless it is well and truly deserved). But often such feelings overwhelm us, and we are simply unable to get rid of them--perhaps for reasons fundamental to human psychology.

Even in (iii)-(v), it is not entirely clear that we are dealing with exactly the right "ought-to-do"'s. In (iii), though we ought not to *have* the flaws and we ought to *take steps* to rid ourselves of them, it is a further step to the flat claim that we *ought to rid ourselves* of them. (Though I believe that stronger claim is part of the Christian doctrine in any case.) The same distinction applies to (iv) and (v). So one might still preserve "'Ought' implies 'can'" by granting both the "ought-not-to-have"'s and the "ought-to-take-steps"'s, but rejecting the stronger "ought-to-rid-oneself"'s (which do sound somewhat odd in cases of established inability).

But let us move on:

(vi) Margolis (1971) points out that conspicuously institutional "ought"'s hold up well against incapabilities. A U.S. Post Office letter carrier ought to complete her appointed rounds, even if a calamitous ice blizzard has made the roads and even sidewalks impassable, and knocked out all telephones as well. A Marine sentry ought to be at his post, even if he has been captured and sedated by terrorists.

(vii) I ought to pay my debts. (White [1975, p.] points out that the very word "ought" derives from "owing" or "owed.") There may be some innocent reason why I cannot pay; perhaps I have been robbed of all my savings. Or there may be a guilty reason, as when I have squandered the savings on ephemera such as Big Macs and cheap whiskey. But in either case, I still ought to pay my debts, even if there are no longer debtors' prisons.

(viii) "Oughts" are sometimes even *highlighted* by the corresponding inabilities, as when a trusted babysitter gets incapably drunk and renders himself unable later to rescue a two-year-old who has fallen into the pool. White (p. 149) makes an aggravated version of this point, observing that if "'ought' implies 'can'," in such a case, we would have a simple way of avoiding what would otherwise be our obligations: we could just prudently, and cynically, render ourselves incapable in advance.

(ix) If I commit an act I know to be wrong, then I ought to repair all the damage I caused. But it may not be possible for me to do this, no matter how painfully I regret my deed.

(x) Finally, there are moral dilemmas, both innocent and guilty. If through no fault of my own I find myself with conflicting obligations, I cannot (entirely) do what I ought--or so it is often argued.[33] As in (vii), the case is clearer when the fault is mine, e.g., as in (ix), a wrongful act of mine may make me liable for reparations to a multiplicity of people, though I simply cannot pay all of them. Certainly I ought to repair all the damage, but I am incapable of doing so.

All these are plausible objections to "'Ought' implies 'can'," but even collectively they are short of decisive. Their main failing, I would contend, is that the relative modalities that respectively drive them remain unanalyzed and so contentious; the "'Ought' implies 'can'" proponent can haggle, and profitably, over "ought-to-be," "ought-to-do," "taking steps to," the "obligation"/"ought" distinction,[34] conditional obligation, the nature of moral dilemmas, and of course "can" itself in all its myriad forms. Nothing less than a proper settling of every such notion within possible-worlds semantics is needed to adjudicate all the apparent counterexamples.[35]

At the same time, I think, an "'Ought' implies 'can'" proponent bears the burden of proof in the face of (i)-(x). And here too our relative-modalities picture is a sure guide. Once the defender has conceded one or more counterexample cases (assuming the opponent's extensive possible-worlds analysis has not succeeded in dismissing the issue utterly), he/she can use the accessibility relation to start working around the residue of counterexamples. That is, (emulating Lewis' (1979) approach to counterfactuals, time, and laws of nature) the defender can survey a large range of clear positive cases and borderline cases, in addition to the conceded counterexamples, and formulate a low-level generalization that captures the clear cases and rules out the counterexamples, while treating borderlines in a reasonable way, using the accessibility relation as the *lingua franca* in which all the different and idiosyncratic modalities are compared to each other.[36]

I make no prediction as to the outcome of this two-staged investigation. I only think that the "'Ought' implies 'can'" issue will remain unresolved until it is carried out.

(vii) *"Why should I be moral?"* This question--that is, its being asked--has puzzled me since I first heard it however many decades ago. The obvious naïve answer to it is: "If that 'should' is a moral 'should', the question either tautologously answers itself or merely seeks to review the moral reasons for what is already stipulated to be the morally right thing. But if 'should' is understood prudentially rather than morally, the answer is that often you 'shouldn't', and so what? Morality, by its very nature, trumps prudence."

Now armed with the picture of relative modalities defended in this chapter, I am still inclined to dismiss the question as stupid or confused. Either the accessibility relation expressed here by "should" coincides with that of the relevant moral "ought," or it does not; and either way, what could be the point of the question?

A defender of the question could try to escape the dilemma by identifying a "should" relation that arguably avoids it. Perhaps that strategy might succeed. But I doubt it. I think the question is simply a bad one as customarily formulated, and ought to be radically rephrased.

What really troubles the people who voice it, I suggest, is something about the nature of *reasons for acting*, viz.: Is it possible, and if so how is it possible, for a person to have a "moral" reason that not only is distinct from any of that person's concurrent prudential or self-regarding reasons, but can override all those reasons? *That* question is both profound and important, and of course a huge literature in moral philosophy has been devoted to it. But it is not well expressed--I would say not expressed at all--by "Why should I be moral?"

In this chapter I have offered a bouquet of (very) late-blooming research projects. Any reader who undertakes one of them and brings it to completion is kindly requested to send me a free copy of the results.

NOTES

[1] When the context fails to supply any very specific cue, a modal assertion is often utterly pointless:

> "And the insurance?" Callaway asked. "When may the beneficiaries expect to have the claim approved?"
> Dora smiled sweetly. "As soon as possible," she said, and shook his hand.

(Lawrence Sanders, *The Seventh Commandment* [New York: Berkley Books, 1992], p. 38.)
[2] Lewis (1973, pp. 6-8) expounds the relativity of modality splendidly, but very much in terms of recognizable philosophical categories.
[3] Even in modern times, grammarians have used a variety of criteria for picking out "modal" verbs as a subclass of auxiliaries. See Joos (1964), Palmer (1965), Ehrman (1966), Lyons (1968), Greenbaum (1969), McCawley (1988, Ch. 8). Everyone seems to agree on at least the following as modal auxiliaries: "can," "could," "may," "might," "must," "have to," "shall," "should," "will," "would," and "ought." "Need to" is only slightly controversial.
[4] In his *Memoirs* (1991, p.13) Kingsley Amis recalls telling his young sons about sex, and ends the story with the sentence, "In no sphere is it truer that it is necessary to say what it is unnecessary to say."

⁵ Max Schulman's *Anyone Got a Match?* (New York: Bantam Books edition, 1965), pp. 192-3. My eye happened to fall on this passage; any reader can find similar examples with little effort.

⁶ I hope there is no longer anyone who thinks that the differences here reflect lexical *ambiguities* in modal words. (Aune [1967] pretty clearly sinned in this regard; Gibbs [1970] is a possible but unconfirmed villain.) Such ambiguities would be monstrous. The point has already been well made by Margolis (1971), by Wertheimer (1972), by White (1975) and by Kratzer (1977). Instead, the accessibility relation is the perfect device for exhibiting the "ambiguities" as pragmatic. (Yet White further argues convincingly for a very general and widespread pragmatic but syntactically marked distinction between "existential" (roughly metaphysical) possibility and "problematic" (roughly, but only roughly, epistemic) possibility.)

Incidentally, both Wertheimer and White are badly undercited, presumably because underread, by modal philosophers and by linguists. Moreover, Wertheimer is undercited by White, and Kratzer cites neither.

⁷ It is of course possible that Boo's whole utterance has derived illocutionary force, and is simply an indirect request; such requests can be performed using "Could you...?" as a semiconventional locution (Morgan [1978]; Lycan [1984, pp. 178-81]. But here I think Boo's utterance can just as well be heard literally.

⁸ If there were, I suppose it would be *staying alive*. But even terminal patients who have not the slightest prospect of staying alive much longer can need things unconnected with staying alive. There are similar counterexamples regarding the nearly universal goals of *being happy* and *having all one's desires fulfilled*. For interesting further discussion, see Stampe (1988).

Also, a need may be ascribed to one person while the relevant end-state resides in another person: "You need to take out the garbage."

⁹ Sorry.

¹⁰ Sorry again. Notice that "need" does not need (s.a.) an animate subject, as is well argued by White (1975, pp. 104-5).

¹¹ Though, unlike "compel," "compulsion" is sometimes used with less than success-grammar; one may have a compulsion to do something and manage to refrain from doing it.

¹² Notice that (8)'s paraphrase in terms of "possible" is far better complementized by "for...to" than by "that": "It is not possible for me to talk any more" is lots better in this context than "It is not possible that I talk any more." Some relative modalities may be syntactically and semantically,

not just pragmatically, pegged to individual subjects. The application of this idea in modal logic was explored by Hilpinen (1969).
[13] Wertheimer (1972, Ch. Three) and White (1975) make excellent starts on a real analysis.
[14] If there are any. I take it to be controversial that there exists any alleged law of nature that is (a) true, (b) strict and exceptionless, *and* (c) indeed physically necessary, in the sense of some event's being even hypothetically ruled out by (because logically incompatible with) a set of basic scientific principles that are themselves strictly true. *Perhaps* "Nothing, relative to any reference frame, travels faster than the speed of light" is such a law.
[15] Noting that a speaker need not be able to specify any one "System" (or set of background conditions or accessibility relation), Wertheimer (1972, p. 92-5) tries to resolve the vagueness by prefixing an existential quantifier: "NP □$_{(whatever)}$ VP" is true iff "there *exists* an *adequate relevant* System," i.e. (p. 95) a System *of the relevant kind.* But the vagueness persists, in "relevant." An hearer may be capable of working out that a particular "must" is moral in kind; but vagueness remains within the category of morality, and the same hearer would in any case have got as far as that category without having to compute an existential quantification.
[16] Interestingly, "unspeakable" cannot be replaced in this context by "unsayable." Indeed it is hard to say, out of context, what the result of that replacement might mean.

It is also interesting to observe that some modal adjectives of this type take intensifiers and/or comparatives: "That glass is very breakable"; "Compound A is more soluble than compound B." A true linguistic Nazi (I myself border on being such a person) might insist that "very breakable" be replaced by "very readily breakable" and "more soluble" by "more readily soluble"; but that seems pedantic, and there would remain the question of why a purely alethic modality would take an adverb like "readily."
[17] Kalb and Kalb (1974, p.400).
[18] All right, just fat.
[19] I follow White (1975, p. 2): "Of many queer philosophical views it is true to say that there is modality in their madness."
[20] Which alleged sense will be impugned in Chs. 9 and 10.
[21] As Unger himself would maintain:

What we properly regard as an accident, or as accidental, does appear to depend on our various interests, as well as upon other things. Thus, even in the most physically deterministic universe imaginable, automobile accidents may occur, and it may be largely accidental that one man, rather than another, is successful in his competitive business enterprise. (p. 159)

[22] Of course some theologians have been unwilling to require that miracles violate natural law. But that liberal view is of no concern to the metaphysics of modality, so I shall not discuss it here.

[23] Lewis (1979) itself offers an account of world-similarity explicitly designed to allow for miracles (though it is aimed specifically at the truth-evaluation of conditional sentences, which may be special). According to that account (p. 472, defended by appeal to a variety of examples),

(1) It is of the first importance to avoid big, widespread, diverse violations of law.
(2) It is of the second importance to maximize the spatio-temporal region throughout which perfect match of particular fact prevails.
(3) It is of the third importance to avoid even small, localized miracles.
(4) It is of little or no importance to secure approximate similarity of particular fact, even in matters that concern us greatly.

But the application of this user's manual to miracle hypotheses is problematic, to say the least. (1) is troublesome--for Lewis' own analysis of miracles, I emphasize, not for his theory of counterfactuals--if we suppose that there could be miracles in Swinburne's sense that *are* big, and collectively widespread and diverse. (Of course, each of those features is also a matter of smooth degree.) (2) is dangerous also, since miracles do typically have massive and long-lasting effects on particular fact. (3) is self-evidently at odds with the similarity of a miraculous world to its corresponding well-behaved world. Only (4) is all right. This would never do; but we should not assume that one and the same similarity relation figures both in Lewis' semantics for counterfactuals and in his philosophical analysis of the concept of miracle.

[24] "Defining" may be too strong, since no formal definition is provided; but Swinburne marks the equivalence by "that is" in the sense of "i.e." (p. 26).

[25] Swinburne's reason for preferring his view (or terminology) over Lewis' is again epistemic, that lawlikeness is more important for theorizing and for prediction than is strict truth.

[26] Agglomerativity allows us to move from " $\Box S$" and " $\Box(S \supset A)$" to " $\Box(S \& (S \supset A))$," and closure warrants the further inference to " $\Box A$." Slote's suggestion that the MMP is motivated by the two formal properties is really only conjecture, but plausible because "...these two properties seem inevitably

to be presupposed in use of the [MMP]...and together ensure its truth" (p. 10). Actually, van Inwagen (1985) denies that his allegiance to the MMP is based on agglomerativity and closure, though he does accept all three; he finds the MMP itself as obvious as is any of the relevant principles.

[27] Actually he speaks of the CS's having no part in the *explanation* of the cause itself, but that seems pointlessly specific.

[28] Van Inwagen has replied to Slote in an unpublished note. He denies that his appeal to the MMP was motivated by any more general formal principles such as agglomeration and closure; he maintains that the MMP simply seems obvious to him when the box is understood in any way that stands opposed to moral responsibility. He explores a few more free-will-related alethic modalities *vis-à-vis* the MMP. One or two of these do fail to support the MMP, but those, van Inwagen points out, are modalities conspicuously congenial to the pernicious Soft Determinist. The dialogue seems at that point to degenerate into burden-tennis.

Kapitan (1991) carries the discussion considerably further.

[29] As John Bigelow has pointed out to me, someone might suggest that van Inwagen's argument is valid but its second premise is false: Since I decided to do A in the normal way and my CS figures in A's etiology ditto, and since to make A true is eo ipso to make B $\supset$ A true, B $\supset$ A is not unreachable from my CS, and that is why the second premise is false. That sounds reasonable so long as we use the phrase "unreachable from my CS" or (as I did put it above) "unaffected by my CS." But van Inwagen will want to go back to the stronger notion of *unalterability*. Although B $\supset$ A does result in part from my CS, it is still unresponsive to it and unalterable by me, for B $\supset$ A would hold anyway, no matter what my CS were like.

[30] Lycan (1987, pp. 117-118) protested against this in a somewhat different way.

[31] As Ziff (1987, p. 1) puts it, "What is peculiar about this view is that it has the air of being simply a report of a clearly perceptible relation between the terms 'can' and 'ought'."

[32] Yes; I have. Can you *believe* that?!

[33] By Lemmon (1962) and Trigg (1971), for two. Other theorists have tried to reconcile the existence of moral dilemmas with "'Ought' implies 'can'" (e.g., Williams [1973] and van Fraassen [1973]); still others have entirely denied the existence of moral dilemmas (McConnell [1978], Conee [1982]).

In regard to the "innocent" versions of (vii) and (x), it may be held that although my obligation remains in force, I am *excused* from fulfilling it, or alternatively (Ziff [1987]) my failure is *mitigated*. These I take to be gentle ways of jettisoning "'Ought' implies 'can'."

[34] Margolis (1971, pp. 59-61); White (1975, Chs. Nine and Ten).

[35] Readers who are still hostile to possible-worlds semantics should concede at least that (a) the relevant distinctions and positive analyses must be made in some vernacular at least as powerful as that of possible-worlds semantics, if such there be, and (b) an adequate possible-worlds analysis would *suffice* to adjudicate "'Ought' implies 'can'."

[36] White (1975, pp. 153-7) offers a subtle defense of "'Ought' implies 'can'." It relies on a distinction "between what is internally and what is externally impossible, that is, between what is impossible in the case stated and what is impossible because of something else" (p. 153); that semantic/pragmatic distinction can easily be cast in terms of accessibility. White goes on somewhat cryptically to illustrate the distinction in terms of scope, e.g., "If I have several jobs which ought to be done this afternoon, then the fact that for some reason I cannot do them does not turn them into jobs which no longer ought to be done this afternoon, though it is necessarily true that jobs which cannot be done this afternoon are not jobs which ought to be done this afternoon" (p. 154).

PART II

MEANING

CHAPTER 9

SEMANTIC COMPETENCE, FUNNY FUNCTORS, AND TRUTH-CONDITIONS

It is often said that a person P knows the meaning of a sentence S if P knows S's truth-conditions, in the sense that given any possible world (or possible situation), P knows whether S is true in that world (or situation). This idea of sentence-meaning corresponds fairly closely to what Frege, Russell, Carnap, and other philosophers have had in mind in speaking of the senses, propositional contents, or "locutionary" meanings of sentences; and, not unnaturally, it has encouraged semanticists such as David Lewis, Robert Stalnaker and Max Cresswell to suggest that sentence-meanings or propositions simply *are* functions from the set of possible worlds onto the truth-values.[1]

In order to say what a meaning *is*, we may first ask what a meaning *does*, and then find something that does that.
    A meaning for a sentence is something that determines the conditions under which the sentence is true or false. It determines the truth-value of the sentence in various possible states of affairs .... (Lewis [1972], p. 22)

The explication of *proposition* given in formal semantics is based on a very homely intuition: when a statement is made, two things go into determining whether it is true or false. First, what did the statement say: what proposition was asserted? Second, what is the world like; does what was said correspond to it? What, we may ask, must a proposition be in order that this simple account be correct? It must be a rule, or a function, taking us from the way the world is into a truth-value. But since our ideas about how the world is change, and since we may wish to consider the statement relative to hypothetical and imaginary situations, we want a function taking not just the actual state of the world, but various possible states of the world into truth-values. (Stalnaker [1972], p. 273)

CHAPTER NINE

Attractive and natural as this idea may be,[2] it occurs to me that there are at least two classes of apparent counterexamples to the claim that knowing a sentence's truth-value in each possible world suffices for knowing its meaning or knowing what proposition it expresses (let us call this claim "the Intension thesis"). These apparent counterexamples have gone nearly unremarked in the literature[3] and it is not immediately obvious how they are to be disposed of. Since they have little relevance to any particularly interesting constructions in natural language, they will not by themselves impede any ongoing work in truth-conditional semantics. But I shall go on to argue that substitutional semantics for English quantifiers occasion a very similar difficulty and are to that extent suspect; I believe that the difficulty has also tacitly contributed to some of the standard confusions and misunderstandings that have attended substitutional semantics.[4]

1. LOGICAL SPACE

Consider the sentence

(A)   Either Elizabeth Taylor's gown has been zipprodted or it hasn't.

You, the reader of this book, do not know the meaning of (A), since you do not know what "zipprodting" is.[5] (Alternatively: You do not know what proposition (A) expresses; you do not know what (A) says. (A) is a meaningful sentence in a dialect of American English, by the way. I know what it means. You can take my word for that.[6]) But you do know (A)'s truth-value in any possible world, because you know in virtue of its form that it is true in all possible worlds. Therefore, knowing a sentence's truth-value through all possible worlds does not suffice for knowing its meaning.

You may say, "But, precisely because I *don't* know what 'zipprodt' means, 'zipprodt' is not a word of *my* idiolect, even if it is a word of yours, therefore it's not surprising that I don't understand (A); I don't speak the language (A) is written in." But this is just to accept the counterexample. If you take my word for the fact that "zipprodt" is a verb of the sort it appears to be (and I never lie about philosophical issues), then you still know that (A) is tautologous and hence that (A) is true in all possible worlds. And yet, you say, you do not understand (A) because it is not even grammatical in your idiolect. This makes matters even worse than before.

Now, in the case of (A) there is at least some relevant truth-conditional knowledge that you lack, namely, knowledge of the satisfaction-conditions of "*x* zipprodts *y*." So the general *idea* of truth- and satisfaction-conditions' serving as the locus of propositional meaning is preserved.[7] This suggests a more substantive reply: What is missing--what you lack *vis-à-vis* (A)--is the ability to *compute* (A)'s truth-value in any world. You cannot start with (A)'s atomic constituents, determine their truth-values on the basis of satisfaction-conditions, and then work up toward (A)'s own truth-value via the recursion recorded in your truth-tables, because you do not know how to tell, of Elizabeth Taylor's gown or of anything else, whether it has been zipprodted.

This reply is not completely adequate. First, of course, it does not block the counterexample to the original Intension thesis, for that thesis said nothing about having to compute a sentence's truth-value from the sentence's primitives on up. What the reply suggests is rather that the Intension thesis may easily be modified by adding this requirement. Notice also that in computing the truth-tables of compound sentences by means of truth-tables, we very frequently do not trouble to start at rock-bottom atomic sentences, if we know independently that the truth-value of one of the constituent sentences would not affect the outcome; and in the case of (A), obviously, we do have independent knowledge of this kind. On the other hand, in the usual sort of shortcut cases it is still true that we are *able* to compute from the primitives on up, even if we do not trouble to perform the computation, and perhaps our having this ability is what is required for us to know the meaning of whatever compound sentence is in question. At any rate, the hypothesis that this ability is required would explain why you do not know the meaning of (A).

In fact, there is independent evidence that some such computational competence is required, and indeed evidence that the meaning of a sentence cannot simply *be* a function from possible worlds to truth-values. As both Lewis and Cresswell point out,[8] intensional equivalence of two sentences (their having the same truth-value in all possible worlds) does not suffice for their being synonymous. For logical equivalence guarantees intensional equivalence but does not suffice for synonymy in the ordinary sense; "2 + 3 = 5" does not mean the same as "37 is prime," despite their being true in all the same possible worlds. One crude but plausible account of the meaning difference is that the two sentences are built up out of different constituent concepts by different compounding procedures. So it is reasonable to suppose that a sentence's meaning, in our propositional sense, is not merely a function from possible worlds to truth-values, but is rather such a function taken together with a procedure for computing it. Two intensionally equivalent sentences can differ in meaning, therefore, if their truth-values relative to a given world are computed by compositionally distinct procedures. Lewis (who favors a categorially based grammar) suggests codifying this idea by identifying sentence-meanings with "semantically interpreted phrase markers minus their terminal nodes; finite ordered trees having at each node a category and an appropriate intension" (p. 31). Cresswell points out that this suggestion would result in some sentences' intensions' not being functions of their component intensions, and goes on (p. 55) to repair this defect formally.[9] For our purposes, we may simply take a sentence-meaning to be the pair consisting of a function from possible worlds to truth-values and a computing procedure for that function.

If we adopt this strategy in order to account for the frequent nonsynonymy of intensionally equivalent sentences, we find that it will clear up our problem about (A) as well. You do not know the meaning of (A) because you lack a computing procedure for (A)'s intension; you lack such a computing procedure because you do not know how to find the set of zipprodted objects in any possible world. For now, then, we can live with the claim that P knows S's meaning if P knows S's truth-value at any possible world on the basis of being able to compute S's truth-value at that world from S's primitives on up. (Call that the

Modified Intension Thesis.) Let us therefore move on to my second sort of counterexample.

2. FREGE'S HORIZONTAL

Consider the following formula of the logical idiom developed in Frege's *Grundgesetze* (Frege [18/1964]).

(B) ——17

According to Frege's semantics, the horizontal "——" is a functor which takes objects as arguments and yields truth-values: when it is applied to a name of the True, the result denotes the True; when it is applied to a name of anything other than the True, the result denotes the False.[10] Thus, since (for Frege) to *be* true (or false) *is* just to denote the True (or the False), the formula "——17 is prime" is true, the formula "——17 is even" is false, and our formula (B) is false as well. Now, as in the case of (A), we know (B)'s truth-value not only in this world but in any possible world: (B) is false in any world in which 17 exists and (according to Frege) truth-valueless in any other world, if any. And yet, as in the case of (A), there seems to be no even faintly nontechnical sense in which you or I know the meaning of (B). In fact, (B) is more drastically problematic than (A), in that it seems (B) simply *does not have* a meaning in anything like the ordinary sense of the word: it expresses no proposition; it has no locutionary content; there is nothing that it says. (B) therefore is another counterexample to the Intension thesis, this time because it has no meaning for anyone to know.

Note that (B's) anomalousness does not depend on 17's being an eternal and perhaps necessary object. "——Gloria Steinem"'s truth-conditions are determined on analogous grounds. Nor does the anomalousness arise from (B)'s own noncontingency;

"——┬──Gloria Steinem
  └── Frege attacked psychologism"

is true at any world in which Frege did attack psychologism and in which Gloria Steinem exists, false at any world in which Frege did not attack psychologism and in which Gloria Steinem exists, and truth-valueless otherwise; but it is equally anomalous.

Nor, more trenchantly, does (B)'s recalcitrance arise from anyone's inability to *compute* (B)'s truth-value relative to any world. (B) is unlike (A) in this respect. And therefore (B) is a counterexample even to our modified thesis that P knows S's meaning if P knows S's truth-value in all possible worlds and can compute S's truth-value in any world from S's primitives on up.

A number of new replies are available:

"Who cares about Frege's horizontal? Frege introduced it only for an obscure mathematical purpose; it plays no serious role in the analysis of any *natural* language, which means it would never come to the professional attention of a semanticist such as Lewis, Stalnaker, or Cresswell. So the semanticists who hold our modified thesis may regard the case of (B) as a 'don't-care'." This response is unsatisfactory, I think, for three reasons: (i) Mere oddity or not, (B) is still a *counterexample* to the modified thesis, and so that thesis would have to be newly modified in any case. The most natural move would be to add the qualification that "S" ranges only over sentences of natural languages, not over the formulas of technical logical theories whose truth-conditions have been determined by arbitrary stipulation. But how could such a qualification be motivated? Certainly it is hard to think of any reason why the meanings of sentences of natural languages would be determined by procedures for computing their truth-values through all possible worlds but the meanings of logical sentences would be determined in some other way (you would think that if anything it would be the other way around).[11] (ii) It cannot be objectionable that Frege simply stipulated the horizontal's peculiar truth-conditions, that those truth-conditions were arbitrarily made up rather than discovered in nature. For it is a constitutive tenet of the truth-conditional method in semantics that sentences of natural languages have *their* truth-conditions only in virtue of bearing determinate connections to formulas of various "underlying" logical idioms whose truth-conditions have been provided just as arbitrarily as have the horizontal's. (iii) As we shall see, the horizontal is not alone: I shall argue that (B)'s anomalousness is mirrored in at least one more familiar construction which figures centrally in the analysis of natural languages and which cannot possibly be regarded as a "don't-care."

"You hear (B) as literally meaningless only because you speak English rather than Grundgesetze. If you did speak Grundgesetze, you would understand (B) in any reasonable sense of 'understand,' and it follows that there *is* something to understand and hence that (B) is meaningful after all."[12] To this I protest that I *do* speak Grundegesetze, at least as well as anyone does. There is no basic lack that I suffer *vis-á-vis* Grundgesetze of the sort that I suffer *vis-á-vis* Urdu or Finnish. At least in principle (I admit I am a little rusty), I know everything there is to know about, or of, Grundgesetze. Therefore, there seems to be no sense in which I speak English rather than Grundgesetze and do not understand Grundgesetze at all.

"You say that (B) expresses no proposition, that there is nothing that (B) says, etc. But (B) has exactly the same truth-condition (exactly the same intension) as the sentence, '17 $\neq$ 17.' So (B) is equivalent to '17 $\neq$ 17' and presumably means that 17 is distinct from 17." The premise is correct: (B) and "17 $\neq$ 17" are intensionally equivalent. But, as has been conceded all around, it does not follow that (B) is synonymous, since their truth-conditions are computed differently (one procedure employing our base clause for "———," the other employing our base clause for "="). Thus, even if (B) did have a meaning, it would not have the same meaning as "17 $\neq$ 17."[13]

"Construing sentence-meanings as functions from possible worlds to truth-values is an elegant and powerful device, and the appeal to truth-conditions has considerably illuminated problems in natural language which have always been considered problems of meaning. So we should simply conclude that (B) is meaningful, in that Frege's semantics assigns (B) a determinate intension, even if (B) looks or sounds funny, and that we do know (B)'s meaning even though it may seem to us that in some colloquial sense we do not and that there is no such thing to know." This is all right so far as it goes; perhaps theory will be better off if we fall in with this adjusted notion of "meaning" as the (low) price of a pleasing systematization.[14] But if we are to adopt the intensional model as our official theory of sentence-meaning (giving Lewis and Stalnaker the word "meaning"), we should still try to explain why we have nihilistic preanalytical feeling about (B). If (B) is not meaningless after all, then what is wrong with it?--less pejoratively, how does it differ from the ordinary sentences of natural languages and from the formulas of standard logical theories as well? What is missing?

One obvious difference between (B) and these other, more familiar items is that (B) lacks a translation into English, our home language.[15] Possibly it is this difference that accounts for our funny feeling about (B). But I doubt it. There are sentences of other natural languages which lack English translations. Suppose you are a native speaker of English, but at some time learn Japanese so well that you may be counted as a genuine bilingual. Suppose further that you are reading a particular sentence of Japanese, and you happen to notice that that sentence is not translatable into English. But no "funny feeling" ensues; you did not suspect that the Japanese sentence is meaningless, because, being a fluent speaker of Japanese, you understand the sentence. You understand it by virtue of being a speaker of Japanese. Why, then, do I not understand (B) (or why do I feel that I do not understand (B)), given that I am a speaker of Grundgesetze? The problem remains.

A second difference between (B) and normal sentences is that (B) cannot be used to make any assertion, so far as I can see.[16] It might be suggested that our intuitive recognition of this defect is the ground of our funny feeling about (B). But it seems clear that to say this is just to put the problem off. We would expect that any expression which had determinate truth-conditions could be used to make an assertion; to have truth-conditions is to represent the world as being a certain way, and so to token a truth-conditionally interpreted sentence in the right tone and circumstances presumably is to assert that the world is as it is represented as being by that sentence. Therefore, (B)'s inability to be used in making assertions is just as badly in need of explanation, and for just the same reason, as is our funny feeling about (B).

I cannot think of any further difference between (B) and normal sentences that would explain these idiosyncrasies. The moral is that if we do affirm that (B) means something, we will still be stuck with the fact that (B) is *somehow* defective as compared to normal sentences and formulas; and (B)'s particular sort of defectiveness, though hard to identify, *feels* semantic in nature. By way of simply labelling the problem, we could say that (B) is semantically *mute*, in that it asserts nothing, has no locutionary content, and cannot be used to make a statement.

So far this need not shock or even faintly disturb the working semanticist, since, as I conceded above, Frege's horizontal does not figure in the analysis of any natural language. But I believe there is a similarly afflicted construction that does figure in linguistic semantics; to it I now turn.

3. SUBSTITUTIONAL QUANTIFICATION

Consider

(C) $(\exists x)(y)$Admires xy, Frege)

(C) is not a misprint. "$(\exists x)$" is a substitutionally interpreted quantifier whose substitution class contains a left-hand parenthesis. "$(\exists x)(\ldots x \ldots)$" is true, in the canonical idiom I have in mind, if there is some lexical item t of that idiom such that the result of replacing "x" in "$(\ldots x \ldots)$" by t is a truth. Now "(y) Admires xy, Frege)"--"(y)" here being a *standard* quantifier--will be turned into a truth if (and only if) a left-hand parenthesis is substituted for "x" and the result, "(y) Admires $(y$, Frege)," is true; thus, (C) is true if everyone admires Frege and false otherwise.[17]

The situation is familiar. You and I have the ability to compute (C)'s truth-value in any possible world. And yet (C) seems to me to be gibberish; though perhaps meaningful or "meaningful," it seemingly cannot be used to make a statement--(C) is "mute." The same range of replies that we ran through as regards (B) is available here, but our rejoinders to the replies will be the same.

Notice that (C)'s persisting anomalousness explains a pedagogical phenomenon with which most of us are familiar: Upon being introduced to the substitution interpretation of the quantifiers, students typically respond, "Oh, I see; substitutionally quantified sentences are *metalinguistic,*" meaning that substitutionally quantified sentences assert things *about* some bits of the object-language in which they occur. To this one responds that, no, they do not assert things about bits of language, any more than "Frege attacked psychologism and Moore attacked idealism" asserts things about bits of language just because the base clause of our truth definition which concerns "and" alludes to the truth-values of that sentence's conjuncts.[18] But the students go away vaguely dissatisfied; they see the point of that response, but still feel that for some reason they have not grasped the purport of the substitutional quantifier. What I am suggesting is that there is a good reason for the students' puzzlement: The students want to know, given that the substitutionally quantified sentence does not say anything about bits of language, what it *does* say, and the reason their puzzlement persists is that the sentence does not say anything.

Perhaps the case of (C) is not quite so clear as that of (B); perhaps we should not be so quick to conclude that (C) is at best semantically mute. For, once the quantifier is understood as I have explained it, there is some temptation to say that (C) *contains* a determinate assertion, namely, the assertion that everyone admires Frege. (One might be similarly tempted to say that "Ugh umph veeblefetzer and everyone admires Frege" contains that assertion.) Perhaps (C) does say at least that everyone admires Frege; let us keep this in mind as a third possibility.

Now, as we all know, no *semanticist* has ever used a substitutional quantifier as powerful as the one I have introduced; no sentence of any natural language would have (C) assigned to it as its logical form. This is because the substitutional quantifiers that semanticists do make use of are intended to capture the roles of quantificational expressions of English or other natural languages, and such expressions (at least superficially) appear to bind pronouns, just as do those quantificational constructions which semanticists construe objectually. Accordingly, semanticists who introduce substitutional quantifiers restrict the substitution classes of those quantifiers to *singular terms* or (in the case of higher-order quantification) to predicates or whole sentences and exclude silly and structurally unimportant lexical items like left-hand parentheses. Therefore, working semanticists, whose substitutional quantifiers are restricted in this entirely natural and reasonable way, will be as undisturbed by the anomalousness of (C) (which will be uninterpretable in *their* canonical idiom) as they are by the anomalousness of (B).

I shall now argue that there is a theoretical reason why the semanticist ought to worry about (C) nevertheless. The argument will be a bit roundabout, but I believe it has some force and calls for a response from any semanticist who continues to appeal to substitutionally interpreted quantifiers as underlying (some) quantificational construction in English.

Let us begin with an uninterpreted formal language. Any such language may be considered and then interpreted in any way one likes, so long as the interpretation yields a truth definition for the language that assigns determinate truth-conditions to each well-formed formula of the language; subject to this last requirement, the interpretation of formal languages is otherwise arbitrary. Now consider an uninterpreted language containing operators which look like quantifiers and which, according to the (syntactic) formation rules for the language, are called "quantifiers." We may propose two distinct interpretations for this language, each of which treats the "quantifiers" substitutionally. The two interpretations are exactly alike, except that one applies only to variables in singular-term positions and accordingly restricts the quantifiers' substitution class to singular terms, while the other is syntactically general and assigns an appropriate substitution class to each of the other syntactic categories as well. Now consider

$$(D) \ (\exists x)(y) \ Admires \ (y,x)$$

I shall argue two points:

and

(1) that (D) on the second of our two substitutional interpretations of the quantifier has exactly the same semantical status (whatever it may be) as (C) has;

(2) that (D) on the first of our two interpetations (that involving the syntactically restricted substitution class) has exactly the same semantical status as (D) on the second interpretation has.

In support of (1), I would point out that, so far as formal semantics is concerned, the fact of the "quantifier"'s "binding" a "variable" that appears in *singular-term* position is totally accidental; our semantics gives singular-term positions no pride of place within our system of syntactic categories. For this reason, though the interpretation itself is entirely clear from a model-theoretic point of view, the "quantifier" is no more intuitively a *quantifier* of any sort than is its cognate in (C). In particular, (D)'s "quantifier" on our second interpretation has nothing whatever to do with existence, even existence very colloquially and fancifully construed. To see this more clearly, note that there is no uniform informal way of paraphrasing the "quantifier" into any even faintly quantificational expression of any better-understood formal or natural language that makes clear sense of (C) as any sort of existential assertion (objectual or otherwise).

It is important not to be misled by the fact that in (D) the "bound variable" happens to replace a singular term, by the orthographic similarity of the "quantifier" to an objectual quantifier, or by (D)'s similarity to sentences of more familiar idioms that we would translate into existential sentences of English; that these similarities are specious is just what is shown by the underlying semantical parity (identity) of (D)'s "quantifier" with (C)'s. It is also important not to be misled by the fact (already acknowledged) that the singling out of the syntactic category of singular terms is not at all arbitrary or "accidental" for serious philosophical and technical purposes such as that of analyzing natural languages. For now I am considering (C) and (D) solely as formulas of a logical idiom to which we are assigning formal semantical interpretations, regardless of what uses these formulas thus interpreted may or may not turn out to have. And the moral of the foregoing argument is that, from *this* abstract point of view, there is no reason at all to think that (C) and (D) on our second interpretation differ in their semantical status or that their "quantifiers" differ in any semantically relevant way.

In support of (2), I would return to the point that I believe Quine (1969a, p. 106) had in mind in reminding us of the kind of construction exemplified in (C). Again: *From the purely formal truth-theoretic point of view,* the restriction of substitutional quantifiers' substitution classes to singular terms is *entirely* arbitrary--not just in that any truth-theoretic interpretation of any logical functor is arbitrary, and not just in that the choice of singular terms from among all the language's syntactic categories is arbitrary, but also in that the idea of making any such choice at all is arbitrary. (The urge to restrict the use of our "quantifier" to singular-term positions is not a mathematical urge; it comes only from philosophers' or linguists' desires to model or parody natural languages.) And it seems therefore that the same sorts of considerations that assimilate (D)'s "quantifier" on our second interpretation to (C)'s "quantifier" also suffice to assimilate (D)'s "quantifier" on our first interpretation to (C)'s. The fact of that interpretation's having restricted the application of (D)'s "quantifier" to singular-term positions is accidental in just the same way as is the fact that (D)'s "variable" on the second interpretation occurs in singular-term position. And, as before, (D)'s "quantifier" even on our first interpretation is no more intuitively a quantifier of any sort than is (C)'s; it has nothing to do with existence in any sense or with any quantificational expression of any more familiar formal or natural language, and so on. Here again, we must resist being misled by the "quantifier"'s superficial similarity to a quantifier etc., or by the nonarbitrariness *for purposes of linguistics* of singling out singular-term positions, for the same reasons as before. There is equally no reason to think that (D)'s semantical status on our first interpretation differs from (D)'s semantical status on our second, or from (C)'s, and there is no reason to think that their "quantifiers" differ in any semantically relevant way.

We concluded above that (C) is semantically mute. Thus, given (1) and (2), we should draw the further conclusion that (D) on either of our substitutional interpretations is semantically mute also, and that any inclination to deny this arises only from (D)'s mendacious orthographic familiarity. (We did consider allowing that (C) asserts that everyone admires Frege, but no analogous option is available here, since (D) does not "contain" any closed sentence.)

Even now we have said nothing to trouble the working semanticist. I have talked only about the logical particles of a couple of nonstandard and useless formal systems. The trouble begins, however, as soon as we turn at last to the analysis of natural languages and try to explain some of their surface constructions in terms of underlying substitutional quantifiers, relating relevant surface expressions to the underlying quantifiers via fairly simple lexicalizing transformations. For example, as we saw in Chapter 1, Marcus (1975-1976) supposes that an expression of English such as "There are things that" in

(E) There are things that don't exist

reflects a substitutionally interpreted quantifier, since it is a plainly quantificational expression but cannot comfortably be regarded as straightforwardly mirroring an objectual quantifier, at least not in the absence of an elaborate Actualist or Concretist metaphysics of possibility. Besides, in a way we do talk as if (E) were made true by the truth of certain substitution-instances of (E)'s matrix: when one tokens (E) in a philosophical discussion one feels an inclination to add, "...*Pegasus* doesn't exist; *Santa Claus* doesn't exist; and so on." But if substitutionally quantified formulas of formal languages are, as I have argued, semantically mute in the sense that they (can be used to) assert nothing, and if whatever transformations derive English sentences from the interpreted formulas of underlying formal languages preserve meaning and all meaning-related properties (as is widely insisted), then on the hypothesis that "There are things that" in (E) reflects a substitutional quantifier, (E) itself is semantically mute. Congenial as that consequence would be to a ferocious antiMeinongian, as a contention about ordinary English it is pretty obviously false. I have already argued in Chapter 1 that Marcus' substitutional proposal will not do, but here is a new difficulty.

The point generalizes quickly to any theory according to which any perfectly meaningful sentence of English reflects an underlying substitutional quantifier--e.g., a sentence containing apparent higher-order quantification ("There are three things I am concerned to deny:...").

The consequence that (E) is mute, taken together with the antecedent plausibility of one or more substitutional-quantifier hypotheses, may make us suspicious of my argument for the muteness of (D). I suspect that some semanticists will simply ignore the argument for that reason. But I think that to do so would be a mistake. The argument *seems* to be sound; therefore, to find a flaw in it would probably teach us something interesting and perhaps important about meaningfulness or about the surrounding semantical and syntactical notions.

4. WHAT IS SEMANTICAL MUTENESS?

We have left section 2's residual question unanswered. What is "semantical muteness"? That is, what is it about a formula such as (B) that makes us feel that that formula asserts nothing and is semantically deficient in some way? Not (B)'s untranslatability into English, as we have seen, or any more basic pragmatic property of (B) that I know of. At present I have very little to offer on this. One feature of Frege's horizontal stands out, however, and it is one that is shared by (C)'s substitutional "quantifier": Each of these "funny functors" is syntactically ambivalent or promiscuous, in that it operates grammatically on arguments of more than one grammatical category. The horizontal applies either to a singular term or to a closed sentence; the "quantifier" applies to an "open sentence" whose "variable" may occur in any grammatical position. (And I argued that the "quantifier" in (D), though by happenstance (on our second interpretation) or by fiat (on our first) it "binds" a "variable" in singular-term position, *might just as well* apply in the same way to a position of any other syntactic

218   CHAPTER NINE

category.) If anything can explain (B)'s and (C)'s felt emptiness, I should say, it must be this property of their contained funny functors.

5. A NEW COUNTEREXAMPLE

I want now to present a more general and philosophically illuminating counterexample, which does not turn on the eccentricities of any unusual and perhaps negligible construction.

Suppose *per impossibile* that God presents me with a logical spaceship, and offers to accompany me on a tour of all the possible worlds there are. (This is a cross between *Star Trek*, the temptation of Jesus, and Virgil leading Dante through the Inferno.) God also carries a placard with some inscrutable writing on it, and assures me that the writing is a sentence of some supernatural tongue. At each world He tells me the sentence's truth-value at that world. I keep a list of all the truth-values *seriatim*, or remember them. Do I know the sentence's meaning? Do I understand it? Hardly.

This case is itself no counterexample to the Modified Intension Thesis, for in it I am not able to compute the sentence's truth-value from the primitives on up. But now we can apply the same principle to a supernatural expression that is a primitive predicate rather than a whole sentence; call the predicate P. God tells me at each world which denizens of that world satisfy P. Now, by applying Mill's methods or by performing some intuitive feat of abstraction to see what all the world-bound extensions have in common, I might then come to know P's meaning, but (it seems to me) I also might well not. And if not, I would not know the meaning of a sentence in which P occurred, despite having the best possible Authority in the matter of truth-conditions.[19]

Let us consider two possible objections. First:[20] If God showed me all the extension of our mysterious predicate P at all the possible worlds there are, then I can form the union of those extensions and give it a name, say, "Fred"; and I understand both the new word "Fred" and expressions of the form "n ε Fred." Now, I also know that for any name N, ⌜N is P⌝ and ⌜N ε Fred⌝ are necessarily equivalent. If I understand ⌜N ε Fred⌝ and know that ⌜N is P⌝ is necessarily equivalent to it (and has no further semantically relevant structure), then surely I understand ⌜N is P⌝ (for what is left to understand?), and so must understand P itself after all.

To this I respond by denying the principle that if I understand an expression E_1 and know of a second expression E_2 that is unstructured and necessarily equivalent to E_1, then I understand E_2. Consider the unstructured sentence "%," which I stipulate to be true at every world. "%" is now necessarily equivalent to every necessary truth. I understand many necessary truths, but I do not understand "%"; I do not know what it says. Similarly, let "*" be an unstructured predicate that everything satisfies at every world. "*" is equivalent to "is red or not red," "is self-identical," "is not both prime and composite," etc, but (intuitively) synonymous with none of them--why would it be synonymous with any one to the exclusion of the others?

Why do I *not* understand P, or "*"? The rhetorical question parenthesized in our first objection is a good one. I have no very secure answer to it, but what I *seem* to lack is the knowledge of what it is about a thing in virtue of which P or "*" applies to that thing (and of what it is about a world that makes "%" true at that world).[21] The notion of a satisfaction-or truth-maker is ill-defined, but seems apposite here. I shall be able to say a bit more about what is missing, in reply to the second objection.

Which is:[22] When I acquire my God-given ability to tell at any world which things in that world satisfy P, do I not emulate the celebrated chicken sexer,[23] who by dint of much training and repetition can tell the sex of a baby chick without knowing how he/she does this? Just as the chicken sexer relies on visual cues to which no one may ever have conscious access, might I not simply acquire the ability to spot instances of P without knowing what features of these items make them P-satisfiers? We do not accuse the chicken sexer of failing to understand "male" or "female" as applied to chicks. Why, then, should we deny that I understand P?

This comparison raises the new question of degrees or *grades* of understanding, which I shall take up in the next section. But there are two more immediate and important replies to be made.

(i) The chicken sexer's subconscious cues are (presumably) neither the genetic and hormonal features of chicks in virtue of which the chicks have the sexes they do, nor the defining features that the chicken sexer may have gleaned from a dictionary or from early exposure to normal English speakers (such as capability of bearing young, and capability of fertilizing a female conspecific). The chicken sexer knows perfectly well what "male" and "female" mean, and there is nothing subconscious about this semantic competence; what is subconscious is rather the *evidence* on which his particular judgments are based. The same is not true of my ability to spot instances of P. It is not that I process the evidence of P-hood subconsciously even though I know what P-hood itself is; it is that I do not even subconsciously know what P-hood is at all, though I do know, on the basis of authority, which things have got it.

(ii) Subconsciously or not, the chicken sexer has a *procedure* (perhaps amounting to an algorithm, perhaps only a very well-trained Connectionist network) for judging the sex of a chick, that keys on features of the particular chick being examined. The chicken sexer exercises a discriminative capacity based essentially on this examination. No such thing holds in the case of my sorting of P-instances. Faced with a thing that in fact satisfies P, I have no way of detecting this unless I can recognize this particular individual and remember that God tagged it as falling within the extension of P. My procedure for spotting P-instances works essentially by authority and not, so to speak, from within.

This all sounds--unpleasantly to my ears--like the neo-Verificationism fomented this decade past at Oxford and elsewhere; but happily it is not that. My complaint about P is not that I have no algorithm for recognizing instances of P in this world, subject to the limitations of my geographical situation and representational and reasoning capacity, for while God is my guide I am omniscient and epistemically unlimited; nothing is "verification-transcendent" for me. My complaint is rather that *even though I am omniscient as regards this world and every other*, I have only the evidence of authority and not a recognition-procedure of my own for spotting instances of P. I

SEMANTIC COMPETENCE 221

do not have inner mechanisms of the sort appropriate for genuine understanding.

For the case of primitive predicates, the modified Intension Thesis collapses into Turing-Test-ism regarding semantic competence; the Thesis counts a subject as understanding a primitive predicate so long as the subject can make the appropriate public discriminations, no matter how, even if the discriminations must be made throughout logical space and not only within our own world (which is only to say that counterfactual as well as actual discriminations are required). But Turing-Test-ism is Behaviorism in the philosophy of mind. I am a Functionalist, not a Behaviorist,[24] and so I find the Turing Test overliberal; I trust you are and do too. Just as a hollow anthropoid shell does not have beliefs, desires and sensations merely because some clever MIT students sit at a console and act as puppeteers by remote-control, making the hollow thing behave just as if it had beliefs, desires and sensations, I do not understand predicate P merely because God has tipped me off as to when to bark out ⌜P!⌝ in a convincing tone.

I conclude that my counterexample stands, and I think this last point about Turing-Test-ism suggests (though certainly does not establish) a moral: It shifts the focus away from the hyperuncountable crystalline array of possible worlds *per se*, and back to the mechanisms operative within the understanding subject, back into the structure of the speaker's competence. Spitting out the right extensions at the right worlds is important only *qua* manifestation of that competence. And so if the competence can be functionally described in its own right, the allusion to possible worlds may prove expendable in principle.[25]

6. GRADES OF UNDERSTANDING

We should admit that there is some force to the behaviorist intuition. For one thing, a person who systematically applies P or any other predicate without fault will rightly be credited with having mastered the community's use of that predicate, irrespective of any further semantical shortcomings on that persons's part.[26] For another, there seem to be cases in which computation of a sentence's truth-value "from the primitives on up" is not after all necessary for what would count as understanding in those cases. It seems to me useful to distinguish at least three different grades of understanding.

I am aware that there is a cyclotron at (say) Cal Tech, and I know what a cyclotron is. Many other people are aware of the same fact, however, without having the faintest idea whether a cyclotron is animal, vegetable or mineral.[27] Do such people understand the word "cyclotron"? Not fully; and not even in the sense of applying it to the right objects in other possible worlds. Yet there is a minimal sense in which they do understand "cyclotron"; they get by in ordinary conversation without provoking linguistic indignation--they use "cyclotron" as a common noun, and know *some* truths incorporating it even though they do not know what a cyclotron is. Let us credit these people with "proto-understanding." (If you find yourself wanting to insist that they do not understand "cyclotron" in any sense at all,[28] and that they are merely cocktail party phonies who *pretend* to understand a word when they really do not, change my term to "pseudo-understanding.")

A second, higher grade of understanding is illustrated by a case of Jay Garfield's (1982). Suppose some of our ancestors spoke what he calls "MegaChinese," a collection of ten thousand unstructured symbols each of which has a truth condition (a distinctive truth-set of worlds). The MegaChinese speakers are "a taciturn, simple lot" and form a primitive society in which basic needs are met but nothing ever happens of any conceptual complexity or interest. (They are very much like Wittgenstein's builders, whose language consists of simple unstructured commands regarding the deployment of various kinds of construction materials. Brian Loar has imagined a similar community whose members use "simple signals."[29]) Now, surely these primitive people understand the symbols of their language, for they grasp both the correct use and the truth conditions of those symbols. Garfield takes this to constitute a counterexample to the converse of the modified Intension Thesis; *computation* of truth condition from the primitives on up is not necessary for understanding.

I find this pretty persuasive, and have just two small points to make in reply. (i) The case is a good deal less straightforward than it might initially appear, in that the *learning* of MegaChinese is mysterious. Our ancestors could not have learned it in the normal Mill's-Methods way, since that way keys on structural elements of sentences and by hypothesis MegaChinese sentences have no structural elements. It seems to me psychologically dubious to suppose that humans might learn each of ten thousand *sentential* items without exploiting any structure internal to those items. (On the other hand, we learn many more than ten thousand primitive predicates without difficulty; and the point is only an empirical one in any case.) (ii) Here as in the case of "%," there is no way of saying that a given sentence of MegaChinese means that p rather than q, where these are logically equivalent but by no means synonymous. So although there surely is a sense in which our primitive people understand sentences of MegaChinese, they do not understand them in the fuller sense of *knowing what they say*.

I conclude that we must distinguish two further grades of understanding. With deliberate tendentiousness, let us call what our ancestors had "behavioral" understanding, and what they did not have "semantical" understanding. Even semantical understanding comes in degrees, n.b., for it is rare that any speaker actually grasps the full extension of a primitive predicate in every detail; even enthusiastic and accomplished birdwatchers may not be fully expert ornithologists. So the distinction between proto- or pseudo-understanding and semantical understanding is not sharp; the two shade off into each other.[30]

Some readers may want to take sides on the question of whether it is behavioral or semantical understanding that is *real* understanding.[31] If I had to choose, I would go for semantical understanding and regard behavioral understanding as a degenerate form, but so far as I can see this is solely a matter of taste and emphasis; there is hardly a further fact of the matter.

7. INDEXICALS

Heidelberger (1982) has raised a related problem concerning our understanding of sentences containing indexicals, and he argues that a purely propositional account of meaning does better at handling this problem than does a truth-conditional account. As an ardent proponent of truth-conditional semantics and at least a temperamental opponent of propositions, I am concerned to rebut Heidelberger's argument and to solve the problem in truth-theoretic terms.

Heidelberger thinks of understanding as the holding of a correct tacit belief as to an expression's meaning; let us fall in with this for purposes of discussion. Thus, let us take (2) to entail the conjunction of (3) and (4), and vice versa.

> (2)   Peter believes correctly that "Je suis paresseux" means "I am lazy."
>
> (3)   "Je suis paresseux" means "I am lazy."
>
> (4)   Peter understands "Je suis paresseux."

Now, the problem is that (3) "contains what appears to be a puzzling use of the first person pronoun" (p. 23). (3) is certainly not equivalent to

> (5)   "Je suis paresseux" means that I am lazy,

for (5), if inscribed by me (WGL) is necessarily equivalent to

> (6) "Je suis paresseux" means that William G. Lycan is lazy,

which is false for almost every token of "Je suis paresseux." (If Peter enters a room and finds a token of it written on a blackboard, he will not understand it as saying that William G. Lycan is lazy. For that matter, if *I* enter the room and find the token, I will not understand it in that way either. Both Peter and I will take it instead of say of whoever inscribed the token that that person is lazy.)

What has truth theory to contribute to this issue? Heidelberger states the truth-condition of "Je suis paresseux" in this way (p. 29):

> (7)   Any occurrence of "Je suis paresseux" [that expresses a proposition] is true if its author is lazy.

But suppose Peter (falsely) believes[32]

(8) "Je suis paresseux" means the author of these words is lazy.

Then surely he will infer (7); but his resulting correct belief of (7) does not suffice for understanding, since by hypothesis he *mis*understands "Je suis paresseux." Heidelberger concludes that "awareness of truth conditions is...nowhere near a sufficient condition of understanding" (p.30).

I have already granted--indeed insisted--that awareness of truth-condition is insufficient for understanding, but I am uncomfortable with Heidelberger's intensifier, "nowhere near," and I reject the contention that indexicals pose a *special* and intractable problem for the modified Intension Thesis; in what remains of this chapter I shall try to refute that contention.

Donald Davidson, leading apostle of truth-theoretic semantics for natural languages, himself pointed out the difficulty that indexical pronouns and other deictic elements create for that program (1967b, pp. 318-20). He offered a solution--roughly, to relativize truth to a speaker and a time--but that solution has been found to be unsatisfactory in each of several ways.[33] Heidelberger has presupposed a different solution, in offering (7) as his interpretation of the truth theorist's account of "Je suis paresseux," and so left it open that yet a third truth-theoretic solution might not succumb to the same objection he put to (7). I have proposed a third solution of my own (Lycan [1984, Ch. 3]); so let us see whether it helps.[34]

Let us adopt the following strategy in writing T-sentences. Target sentences[35] will be considered true or false only relative to contexts of utterance. I shall suppose that lurking in the background is a valuation function α that assigns denotata to indexical terms in context. Thus, an adequate pragmatics would tell us how α is computed. One rule of a correct pragmatics for French would say that "Je" denotes in context C whomever is uttering it in C (unless it occurs inside direct quotation, etc.[36]).

226   CHAPTER NINE

According to this method, a T-sentence directed upon "Je suis paresseux" would be

(9)   "Je suis paresseux" is true in C iff α("Je,"C) is lazy.

Someone who knew that α("Je," C*) = Jones, where C* is a context in which Peter has come upon a token of "Je suis paresseux," would be able to go on and derive the "purer" T-sentence

(10)   "Je suis paresseux" is true in C* [or: Peter's token of "Je suis paresseux" is true] iff Jones is lazy.

But Peter does not know that.

(9) cannot be substituted into (2) *salva propositione*. But no truth-condition theorist has ever maintained either that a T-sentence could be so substituted into s sentence like (2) or that knowing an extensional T-sentence suffices for understanding or knowing the meaning of the target sentence. (In addition to Heidelberger's objection to the latter thesis, there is the fact that someone might know the truth-values of both sides of a T-sentence and infer the truth-value of the T-sentence itself, without understanding the target sentence.) Indeed, Davidson insists repeatedly that it is not a T-sentence alone, but the T-sentence along with its derivation and in the context of its whole containing truth theory that gives the meaning of the target sentence upon which it is directed. Accordingly, one must not only know (or believe correctly) that "Je suis paresseux" is true in C iff α("Je," C) is lazy, but believe it on the basis of at least an implicit Tarskian derivation from the usual sorts of base and recursive clauses; one must begin by knowing that "paresseux" is satisfied by exactly those things that are lazy, and so on.[37]

It is hard to state a Davidsonian counterpart for (3). He offers no analysis of "x means y," and can be understood in "Truth and Meaning" as simply eliminating it rather than trying to explicate it. If pressed, he might try modelling and analysis of "x means y" on his treatment of indirect quotation (see Davidson [1968]). But a truth-conditional account of "Je suis paresseux"'s meaning would certainly be close to what Peter tacitly knows: that "Je suis paresseux" is true in C iff α("Je," C) is lazy and that this is because... (here follows the derivation). Thus, a Davidsonian counterpart of (2) might well entail a Davidsonian counterpart of (3), and would entail something as near as matters to (4).[38]

Davidsonian's own extension-hugging approach creates many problems. It is hard to see how a truth theorist could succeed in explicating counterfactuals and such without somehow intensionalizing the whole operation, presumably by dint of introducing possible worlds. So let us briefly consider and alternative, model-theoretic attempt (I suppress most of the usual parameters):

(11) "Je suis paresseux" is true at $<C,w>$ iff α("Je," C) is lazy at w.

C is our context of utterance at this world, as before while w is the world relative to which our target sentence is being evaluated. And as before, someone who knew that α("Je," C*) = Jones could derive

(12) "Je suis paresseux" is true at C* at w (or Peter's token of "Je suis paresseux" is true at w) iff Jones is lazy at w.

But Peter does not know that.

One might think that intensionalizing our T-sentences by relativizing truth and satisfaction to worlds would obviate Davidson's motive for requiring knowledge of the T-sentences' derivations, since objections to (7) that are based on the extensionality of the biconditional no longer go through. This is almost right. If we substitute (12) into (2), we get

(13) Peter believes correctly that "Je suis paresseux" is true in C* at w iff α("Je," C*) is lazy at w.

(13) arguably entails (3) and is entailed jointly by (3) and (4). (13) fails to entail (4), but not for any reason essentially involving

indexicals--only for the anti-behaviorist reason I have defended in section 5. I conclude that there is no *further* problem about indexicals of the sort Heidelberger had in mind.

NOTES

[1] See Lewis (1972, sec. III), Stalnaker (1972), and Cresswell (1973, 1985). Hintikka (1969b) had previously indicated sympathy with this approach, as had Montague (196/1974, 1974). My impression is that the idea has its roots in the *Tractatus*, by way of Carnap (1947/1956) and Kaplan (1964), though perhaps it goes farther back. Donald Davidson also has done much to encourage the view that knowing truth-conditions suffices for knowing meaning (see particularly Davidson [1967b]), though he has no use for the possible-worlds apparatus.

As we shall see, it will be necessary to refine this intensional account of sentence-meanings a bit, due to our actually individuating meanings more finely than we do functions from possible worlds to truth-values. I shall take account of this in due course. However, I shall neglect the further refinements that are needed in order to handle sentences containing indexical or deictic elements (see the aforementioned papers by Lewis, Stalnaker, and Montague, as well as Cresswell [1973]), since deixis does not affect the main points I wish to make in this chapter. We shall return briefly to deixis in section 7 below.

[2] Lycan (1984) defended this thesis as vigorously as I could manage.

[3] But see Heidelberger (1980), and especially van Inwagen (1981), who independently developed an argument similar to mine.

[4] Some of these misunderstandings are pointed out and corrected by Dunn and Belnap (1968), and some others by Kripke (1976).

[5] Unless you are Max Cresswell; see note 7.

[6] For teaching me the concept I am indebted to my cousin, Joan Shepard.

[7] Cresswell put the point this way in conversation. In gratitude and a moment of weakness, I told him what "zipprodt" means.

[8] Lewis (1972, sec. V); Cresswell (1973, pp. 45ff. Stalnaker (1976b, 1984), however, is inclined to disagree.

[9] For his more recent views, see Cresswell (1985).

[10] Frege (18/1964, Ch. 5); see also Frege (18/1952, p. 34). Frege's use of the horizontal is discussed at length by Heck and Lycan (1979).

[11] It certainly does not matter that Frege understood truth and falsity as relations between sentences and objects called "the True" and "the False." We could easily define the horizontal's truth-conditions as he does without taking truth-values to be objects.

[12] Cresswell took this line in a discussion of Cresswell (1978).

[13] As a further way of seeing how essential computing-procedure is to meaning, consider a simple (and silly) formal language whose lexicon contains just two expressions. "%" and "#", each of which is a wff. (Nothing else is a wff.) As an interpretation I stipulate that "%" is true in any world and that "#" is true at no world. "%" and "#" surely are counterexamples to the Intension thesis, intuitively because we do not know what it is about a world w that *makes* "%" true in w or "#" false in w.

[14] In general I am willing to fall in with *anyone's* adjusted notion of "meaning." Philosophers' and linguists' uses of the term "meaning" are irreparably variegated, I believe, and thoroughly tainted with the respective theories of quite different linguistic phenomena that have gone under the rubric of "theories of meaning." My point is not merely that theorists' uses of "meaning" are theory-laden; it is that "meaning" has figured as a label, now in theories of truth-conditions, now in theories of communication, now in theories of inference and inferential behavior, now in theories of speech acts, now in theories of syntactic structure, and so on, and so on, and that all these sorts of theories are by and large compatible theories *of* quite distinct subject-matters or ranges of phenomena (here I follow Harman [1968]). In philosophy and in linguistics, the term "meaning" simply has no standard use at present; it floats freely on the surface of semiotic study as a whole, anchoring nowhere unless at someone's terminological whim.

[15] I say this despite the fact that a number of commentators on Frege have offered English glosses or paraphrases of the horizontal. That no such paraphrase can work is argued in Heck and Lycan (1979).

[16] In Frege's system, (B) may be acceptably prefixed by his vertical judgment-stroke. The resulting expression, "—17," has the superficial form of an assertion, but still would not be recognized as such in any "total speech situation" even among speakers of Grundgesetze. I know of no place in which Frege ever commented on this pocket anomaly.

[17] Quine (1969a) points out the syntactic catholicity of the substitution interpretation; he attributes the point in turn to Lesniewski.

It should be noted that a left-hand parenthesis is the *only* lexical item of any familiar logical idiom that could grammatically replace "x" in (C). A Tarskian base clause for my substitutional quantifier would therefore have to be written with some care. We might simply stipulate that "(" is the sole member of the quantifier's substitution class; in this case the single substitution that would verify an existential quantification would verify the corresponding universal one as well. Or we might write a more general interpretation for the quantifier that would apply to any syntactic position but relativize the substitution class to syntactic category (in this case some of the relativized substitution classes might have to be restricted in order to avoid paradox). A substitutionally interpreted "parenthesis" quantifier is unquestionably a stupid and totally useless device that holds no interest whatever for any working semanticist or logician. My claim is only that a *coherent* truth-conditional interpretation could be written for it. It is crucial to see that my argument in no way depends on any assumption stronger than this minimal observation. (I am indebted to Robert Kraut for insisting on this clarification.)

[18] This point is made neatly by Dunn and Belnap (1968).

[19] Max Cresswell has pointed out, quite rightly, that this counterexample does nothing to impugn the bare metaphysical identification of sentence meanings with sets of worlds. That a device A computes a function while a second device B gets the function right in extension just by luck does not entail that A and B compute different functions.

[20] This objection is adapted from some remarks of David Lewis', in correspondence, but he is not responsible for any misuse I may be making of those remarks here.

[21] Ziff (1972, pp. 17-20) explicates linguistic understanding as "analytical data processing of some sort." Ziff maintains that "only that which is composite, complex, and thus capable of analysis, is capable of being understood." It *seems* to follow from this that what is wrong with our unstructured items P, "%" and "*" is just that they are unstructured. But something is wrong here, for there are (lots and lots of) ordinary primitive predicates of English that we do understand despite their lack of semantic structure.

For a criticism of Ziff on this point and an illuminating general discussion of understanding *tout court*, see Rosenberg (1981).

[22] This second objection is similarly adapted from some remarks of Murray Kiteley's, ditto, ditto fn 20. I am grateful to Lewis, to Kiteley, and to Steven Boër, Max Cresswell, Jay Garfield, Michael Hand and Geørbert Krauschumm for extensive and helpful discussion of this issue.

[23] I was first introduced to this admirable character by Armstrong (1968, pp. 114-115).

SEMANTIC COMPETENCE   231

[24] On the difference as it relates to the Turing Test, see Block (1981). For my own version of functionalism, see Lycan (1987, 1988).
[25] This is not to say that a creature's functional organization determines the truth-conditions of that creature's sentences, which it manifestly does not. See Lycan (1984, Ch. 10).
[26] He or she might be compared to the actual case of a color-blind person to whom red and green are indistinguishable but who has learned to use the terms "red" and "green" appropriately on any normal occasion, using various subtle cues. (Wholly blind people, in fact, have been known to pass in restricted contexts for sighted.) Does such a person *understand* "red" and "green"? I want to say, yes and no; read on.
     There is one difference between our color-blindness victim and wielders of predicate P: There are behavioral tests, albeit slightly contrived ones, that would succeed in picking the former out of a crowd. But the color-blindness victim is not thereby revealed to be an externally animated hollow shell; his or her color judgments are the result of a good deal of autonomous "internal processing that largely overlaps a normal person's.
[27] I owe this example to Carl Ginet (1975, pp. 24-25), by way of my colleague George Schlesinger. Ginet is concerned with the connection between understanding and *being confident*.
[28] Stephen Stich (1979) seems to take this hard-hearted line against a helpless puppy, Fido, denying that Fido has the concept of a *bone*, in part on the grounds that Fido is ignorant of a bone's role in skeletal structure.
[29] Loar (1976) uses the example to make a more ambitious point against Donald Davidson's theory.
[30] Cf. Dennett (1969, p.183):

What are the conditions that would suffice to show that a child understood his own statement: "Daddy is a doctor'? Must the child be able to produce paraphrases, or expand on the subject by saying his father cures sick people? Or is it enough if the child knows that Daddy's being a doctor precludes his being a butcher, a baker, a candlestick maker? Does the child know what a doctor is if he lacks the concept of a fake doctor, a quack, an unlicensed practitioner? Surely the child's understanding of what it is to be a doctor (as well as what it is to be a father, etc.) will grow through the years, and hence his understanding of the sentence "Daddy is a doctor" will grow.

Rosenberg (1981) offers a general characterization of understanding that explains why there is no fixed cutoff point in a case of this sort.

CHAPTER NINE

[31] Loar (1976, p.145) seems to want to:

A person may authoritatively be informed as to what a certain sentence of an otherwise unknown language means, without knowing any facts of its structure-in that sense he knows its meaning, and that is how I shall use the notion. Of course, one might want to say that such a person does not *fully* know its meaning; but I would rather say what he does not know are further facts about its semantical structure; facts as to *how* its language assigns it its meaning.

[32] Falsely, because the sentence "Je suis paresseux" does not mean anything about authorship; it is true, for example, at worlds in which no one has ever written or inscribe anything-indeed, at worlds in which there is no language.

[33] For an excellent discussion see Burge (1974). I criticize Burge's own proposal and put forward a new one (Lycan, 1984, Ch. 3).

[34] It does.

[35] This is my term for a sentence of the object-language under semantical examination, quoted on the left-hand side of a T-sentence.

[36] Actually there is a great deal more to that "unless." The common assumption that "I" always designates its utterer is importantly false; see Nunberg (1993).

[37] An objection similar to the foregoing one could be raised even against this requirement; Someone might know that facta about "paresseux" indirectly, without actually knowing what "paressuex" means; cf "renate" and "cordate." One must actually know more about "paresseux"--its intension in some sense, or perhaps its reference-shifting habits under pressure from nonextensional opperators. Davidson sweeps this under the rug, but it is implicit in his view; one would not know all the truth-conditional properties of "paresseux" unless one knew its semantical contribution from within (e.g.) counterfactual contexts (likewise for "renate" and "cordate").

[38] The discussion so far reminds us of the distinction between behavioral and semantical understanding. If someone knew (3), but knew it only because he had been told it by a reliable informant and did not know what "Je" or "suis" or "paresseux" meant (perhaps he thinks that "je" means *lazy*, "suis" means *I* and "parasseux" means *am*, French being an OSV language), that person would have behavioral but not semantical understanding. It is semantical understanding that is captured (almost) by Davidson's account.

CHAPTER 10

LOGICAL CONSTANTS AND THE GLORY OF TRUTH-CONDITIONAL SEMANTICS

This chapter commemorates "The Logical Form of Action Sentences" (Davidson [1967a]). In so doing, it also celebrates the logical forms of action sentences. The latter are still with us, since action sentences themselves are; but for linguistic semantics generally, the magnificent promise of Davidson's classic article has never been fulfilled. My purpose here is to explore one reason why that is so.

1. DAVIDSON'S WONDERFUL PROJECT

According to (loosely) Davidsonian semantic theory,[1] the core meaning of a sentence--its propositional or locutionary content as recorded in indirect discourse by a "that"-clause--is that sentence's truth-condition. The sentence's truth-condition is determined by the meanings of the sentence's smallest meaningful parts together with their grammatical mode of composition, and it is best represented by a formula of some explicitly truth-defined logical system acting as a canonical idiom. Such a formula wears its own truth-condition on its sleeve, in that its truth-condition is computable on the basis of the usual Tarskian set of valuations for the atomic elements of that system plus a set of recursive rules that project the semantic values of a formula's elements through truth-functional and other syntactic compounding into a truth-condition for the formula as a whole.

So we have our original natural-language sentence, endowed with its truth-condition as represented by its associated formula--call that explicitly truth-defined formula the sentence's semantic representation (SR). By means of its assigned SR, the target sentence's logical and semantical features are predicted, for logical relations are defined in the usual way over the formulas of the canonical idiom. In

233

this way too, logical anomalies are resolved and target sentences' semantically puzzling features are explained, just as Russell imagined at the time he fashioned the Theory of Descriptions. Target sentences simply inherit their perceived semantical features, such as entailments, from the formal properties of their SRs.

To "do semantics" is to assign explicit truth-conditions to sentences. A semantic theorist investigates the semantics of a particular natural language by associating canonical SRs with the sentences of the language, in such a way as to illuminate semantic structure consonantly with what is known of the sentence's syntactic structure. A semantics will have, or should have, testable consequences: it will predict ambiguities, synonymies, anomalies, logical implications and the like. Capturing implications is a main goal, perhaps the main goal, of the enterprise.

Davidson did it (accomplished a capturing of implications) in the most dramatic possible way, for the case of action verbs and their most typical modifiers. Concentrating on the demodification inference (from "Jones buttered the toast slowly, deliberately, in the bathroom, with a knife, at midnight" to "Jones buttered the toast slowly, with a knife, at midnight," "Jones buttered the toast deliberately, in the bathroom," "Jones buttered the toast," and the like), Davidson hypothesized that the modified verbs are really existential quantifications over actions construed as concrete events, and that their modifiers are really conjoined predications of the corresponding bound event variables. Thus: The sentence "Jones buttered the toast" would be represented as

(1) $(\exists e)$BUTTERING$(e,$ Jones, the toast$)$.

("There occurred a buttering, by Jones, of the toast"), or more elaborately, picking up further obvious implications, as

(1*) $(\exists e)($BUTTERING(e) & PROTAGONIST$($Jones$, e)$ & VICTIM$($the toast$, e))$.

Modifiers are just further conjoined predications, which explains why they may be stripped singly or in groups:

(2) (∃e)(BUTTERING(e, Jones, the toast) & SLOW(e) & DELIBERATE(e) & OCCURRED-IN(e, the bathroom) & (∃y)(KNIFE(y) & DONE-WITH(e,y)) & OCCURRED-AT(e, midnight)).

So demodification is at bottom just ampersand-elimination, the trivial derivation from a conjunction of one of its conjuncts. A puzzling and potentially troublesome felt implication is reduced to a clean, extensional, canonical, indeed classical form of inference. And so, captured.

But the notion of "capturing" involved here needs looking into.

2. "MEANING POSTULATES"

The sentence "Mort is incredibly lucky and either an idiot or a genius" is felt nonnegotiably to imply "Mort is incredibly lucky." The sentence "Mort is a bachelor" is felt nonnegotiably to imply "Mort is unmarried." But there is a difference. The former implication is thought to hold in virtue of the target sentence's "logical structure," while the latter is only "lexical," a matter of individual word meaning, viz., of the meanings of "bachelor" and of "unmarried" (or perhaps "married" as modified by "un-"). I do not know how strongly this structural-*vs.*-lexical difference is felt by ordinary English speakers, but it impresses itself on logicians and philosophers; notice that any professional would casually grant that "Mort is incredibly lucky and either an idiot or a genius" *entails* "Mort is incredibly lucky," outright, while the felt implication of "Mort is unmarried" by "Mort is a bachelor" is not so easily described as *entailment* even though it is apodeictic and noncancellable. Real entailment is a matter of syntactic/semantic structure, while implications that hold in virtue of simple word meaning are only halfheartedly felt to deserve the title of "entailment," even when they seem equally apodeictic and noncancellable.

CHAPTER TEN

In a linguistic theory built along Davidsonian lines, many felt implications are captured or codified structurally, but many others, lexical ones, are left over. The theory must deal with them in some different way. The most popular way is the introduction of "meaning postulates," or as Harman (1972b) once called them, "nonlogical axioms"; thus:

MEANING POSTULATES

$(x)(BACHELOR(x) \rightarrow ADULT(x))$.
$(x)(BACHELOR(x) \rightarrow \sim MARRIED(x))$.
$(x)(BACHELOR(x) \rightarrow MALE(x))$.

Meaning postulates can themselves employ forthrightly logical operators such as negation.

Now, notoriously, W.V. Quine has taken a dim view of "meaning postulates" as originally introduced by Carnap. Simply by writing the heading "Meaning Postulates" above the list of generalizations that have been claimed to be analytic, one does nothing to explicate analyticity or synonymy. Nor, in a linguistic theory, does one go anywhere toward explaining a felt implication by writing "Meaning Postulate" above a representation of that implication. One may thereby *record one's impression that* two or more concepts are connected in a logical/linguistic rather than an empirical manner, but such impressions are of no theoretical significance unless themselves explained.

Quine of course denies that there is any theoretically substantive difference between generalizations that hold in virtue of word meaning and those that are simply empirical correlations in the world. Thus he seeks no substitute for "meaning postulates," but would advise simply scrapping the idea of "lexical semantics" even if he were to tolerate linguistic semantics more generally.

LOGICAL CONSTANTS

I am sympathetic to Quine's unsympathetic view. But I am now afraid it will not do as it stands. For the distinction between mere "meaning postulates" and bonafide logical truths proves to be deeply suspect.

3. "LOGICAL CONSTANTS"

A "logical truth" is supposed to be a sentence or formula that is true under any admissible reinterpretation of its nonlogical terms. Nonlogical terms, that is, as opposed to *logical constants*. Logical constants are thus presupposed to differ in kind from ordinary morphemes. But that presupposition is not at all obviously correct. Various authors have encountered difficulty in trying to delineate the class of logical constants in any principled way (e.g., see Peacocke [1976] and McCarthy [1981]). Worse, as has not been generally noticed, mere lexical implications seem to *shade off* into clearly logical entailments by fairly smooth degrees. Consider:

x is a bachelor $\rightarrow$ x is unmarried.

x has a daughter $\rightarrow$ x is a parent.

x is red $\rightarrow$ x is colored.

x is a closed curve in a plane $\rightarrow$ x has an area.

x is a hexagon $\rightarrow$ x is a regular polygon.

x Ved *Adv*ly $\rightarrow$ x Ved.

x has been un Ved $\rightarrow$ x was Ved.

x is very *Adj* $\rightarrow$ x is *Adj*.

x is now F $\rightarrow$ It will be the case that x was F.

x knows that $P \rightarrow P$.

Necessarily $P \rightarrow P$.

x is greater than y and y is greater than $z \rightarrow y$ is greater than z.

Most Fs are $G \rightarrow$ Many Fs are G; Many Fs are $G \rightarrow$ Some Fs are G; Some Fs are $G \rightarrow$ At least one F is G.

There are n Fs $\rightarrow$ There are at least $n - m$ Fs [for integers n and m, $0 \leq m \leq n - 1$].

x is identical with $y \rightarrow y$ is identical with x; x is identical with y and y is identical with $z \rightarrow x$ is identical with z.

All Fs are $G \rightarrow$ Some Fs are G.

There is an F that Vs every $G \rightarrow$ Every G is V'd by some F.

R only if either R and Q or R and not $P \rightarrow$ If P, then Q.

P and $Q \rightarrow P$.

The foregoing examples are listed *roughly* in increasing order of "logicalness." There is plenty of room for dispute about exact placement of the examples. (Is "greater than" really more structural or logical--i.e., less purely lexical--than "necessarily"?) And there would be even more dispute about other examples--e.g., where do various odd sorts of modifiers fit in? But the general moral is clear: In no such good healthy list of examples can any great obtrusive break be seen, between the merely lexical and the genuinely "structural." Logicalness as opposed to lexicalness seems to be a matter of degree or at best of grade.

LOGICAL CONSTANTS 239

The point was made years ago by Max Cresswell, specifically in regard to Davidson's demodification project but also addressing some of Davidson's nonclassical, antiextensionalist opponents in the ensuing literature:

> All these authors [Davidson (1967a), Montague (1970), Parsons (1972), Lewis (1972), and Clark (1970)] seem to have at the back of their minds that (a) it is possible, and (b) it is desirable, to make explicit some or all of the true entailments between English sentences. I am pretty sure that it is not possible to do it for all true entailments and I see little theoretical interest in doing it only for some. The sentence **this is red** entails **this is coloured** because the meaning of these sentences in English is such that in any possible world if it is true to say **this is red** it is also true to say **this is coloured**. Whether an English speaker has come to learn the meaning of **red** by building it up from meaning blocks which include the meaning of **coloured** is of no concern to the logician. (Cresswell [1974, p. 470])

In establishing the format of his own great semantical work, *Logics and Languages* (1973), Cresswell declined to distinguish any set of "logical" words of English from ordinary morphemes. (Though somewhat incongruously he did treat a word of *non-* or pre-English as purely logical: the intensional abstraction operator λ.[2] On Cresswell's theory, λ mediates countless felt implications, but it is grammatically deleted from every sentence everywhere and, oddly, has no English lexical realization at all--a truly modest hero. On this understanding, Cresswell should be read not as challenging the "logical"/"nonlogical" distinction, but simply as maintaining that in fact the English lexicon happens to contain only nonlogical terms, English's one logical constant being systematically suppressed by syntax.)

The Cresswell passage quoted above is striking in that it opposes a realist view of the logical/lexical distinction only with a psychological claim about learning. Evidently Cresswell sees the prevailing conception of a genuinely logical entailment as requiring something to do with a psychological process of building up concepts. But what? One might suppose that an entailment obtains when the entailing proposition requires the entailed proposition in the psychological order of learning. But that criterion would let in any lexical implication that in fact satisfies the psychological condition of *de facto* learning-theoretic priority; it has nothing to do with logic. A stronger criterion would be one according to which entailment requires psychologically real lexical decomposition, sychronic rather than developmental; but that criterion too could well let in implications arising from *purely* lexical decompositions, e.g., as of "x kills y" into "x causes: y die." The most obvious next stronger requirement is that genuine entailments should be explicated in terms of psychologically real lexical decomposition plus genuinely logical constants; e.g., "This is red" entails "This is coloured" only if psychologically the English "red" decomposes into (hence one can learn "red" only by first learning) both "coloured" and the ampersand and in effect "red" abbreviates the concatenation of "coloured," the ampersand, and some third concept that somehow determines the determinable--but that criterion is (at best) unlikely to be satisfied, and is circular to boot.

I suspect that what Cresswell meant to attribute to his opponents is after all the second of the foregoing three criteria, the idea being that if an implication really is authorized by a psychologically real lexical decomposition, then that implication is as genuine a case of entailment as anyone might wish, and other "implications" that do not meet the psychological standard simply do not qualify. On this exegesis, Cresswell reacted with a logician's quite proper instinct to distinguish a logical concept from all matters of psychology. Proponents of what the Generative Semanticists of the late 1960s called "Natural Logic" would balk at this, and we should all agree that a real psychological distinction is a genuine distinction and moreover one to be scrupulously respected by linguistic semantics, but logic does not automatically include all of linguistic semantics, and I would agree with Cresswell (on the present interpretation) that "psychological entailment" of this sort is not *eo ipso* logical entailment in anything like the sense usually attached to that term.[3]

There is a halfway house. We can read "The Logical Form of Action Sentences" anachronistically as an attempt to save the "logical"/"lexical" distinction by showing that intermediate or dubious cases of logical implication reduce to bonafide entailment when real logical form is ultimately revealed, "real" logical form involving only something like the traditional logical operators. And a psychological-reality claim for that revelation would strongly support a special logical status for the relevant class of fundamental operators. Thus there would be a middle ground between the obviously lexical and the obviously logical locutions of English--those which are not logical on the surface but which semantically decompose into fundamental logical operators and nothing but. That middle ground is Davidson's strongest hope.

4. THE UNDERLYING MAGNITUDE

If logicalness as opposed to lexicalness is a matter of degree, what more basic magnitude controls that degree? *Generality of subject-matter* is the obvious candidate. "Bachelor" applies to very few individuals, cosmically speaking, while "and" is useful coinage anywhere. "Know" is restricted to sapient subjects, while "greater than" applies to anything that comes in degrees but to nothing else. I think Peacocke (1976) had the right idea in suggesting that "topic-neutrality" is intuitively the key desideratum: Logic is supposed to be abstract and utterly catholic, in no way attached to any particular topic of conversation. Does that guiding idea not mark off some locutions from the rest?

Peacocke and then McCarthy (1981) maintained that it does, and both authors have constructed formal devices intended to implement the idea of topic-neutrality. But intuitively, once one starts thinking in terms of a continuum of cases, topic-neutrality itself in effect comes in degrees. The locutions featured in my list of examples differ in the degree to which they constrain the ontology of the expressions to which they apply, but only for the case of truth-functional connectives is it *strictly* right to say that there is no constraint whatever. Adverbial demodification is about human or nonhuman agents. Set theory is at least about sets. Quantification theory is at least about a domain of individuals and either sets or properties defined on that domain. One might for this reason insist that only the truth-functional connectives are genuinely logical constants, and not even the quantifiers qualify; I would accept that as a well-motivated stipulation, but only as a stipulation. (N.b., agreement with Cresswell's continuum thesis does not rule out the possibility of defining sharp distinctions near the upper end of the continuum of "logicalness." For example, one might accept a cleaned-up version of Peacocke's distinction having to do with *de re a priori* knowledge, as a genuine distinction and even as providing a good thing to mean for some purposes by "logical constant," without committing oneself to the claim that as a matter of fact, all and only expressions meeting the Peacocke condition are true logical constants.)

There is a possible competing candidate for the magnitude underlying degrees of logicalness, suggested by Quine (1986a). At one point in his chapter on logical truth he suggests tying the logical/lexical distinction to the grammatical notion of a particle; logical constants are particles rather than members of what we might call the lexicon at large. Particles are distinguished from ordinary lexical items in virtue of their falling into small grammatical categories whose membership is fixed. By contrast, lexical categories are indefinite in their membership, typically infinite owing to the recursive compounding of complex expressions.[4] Now, this Quinean view could naturally survive the rejection of an absolute "logical"/"lexical" distinction, for the size of a grammatical category is a matter of degree as well. Thus it might be suggested that "logicalness" is at bottom a matter of that size, i.e., of how few other morphemes there are of the same grammatical type.

I think it is clear that that suggestion will not do. For the actual size of a grammatical category's membership is a highly contingent fact about a given natural language, and it is not *a priori* related to the logical status of any of the expressions that are the category's members. Moreover, if it is iterative compounding that makes the categories of (e.g.) singular terms and common nouns indefinitely large and open-ended, the same applies to the category of binary truth-functional connectives, for English contains nonfinitely many well-formed complex connectives, even though only a few of them are ever actually used and even though all but sixteen of them are logically equivalent to simpler ones. Thus we had better stick to degrees of generality or topic-neutrality after all.

LOGICAL CONSTANTS

5. MEANING POSTULATES REINSTATED

If I am right in agreeing with Cresswell that the "logical"/"lexical" distinction is one of degree rather than one of kind, that in turn impugns the distinction between the official truth-rules that define logical operators (the operators' recursion clauses in a Tarskian truth definition for the containing language) and the measly, pathetic "meaning postulates" scribbled down *ad hoc* to "explain" purely lexical implications. What then becomes of Quinean skepticism about the latter?

Quine's view is that while the recursion clauses in a logical theory may be said to play an explanatory or at least substantive systematizing role in regimenting the sentences of a natural language and exhibiting patterns of valid inference, the heading "meaning postulates" is *flatus vocis*, an empty and pointless gesture of no explanatory value whatever. But once we see logicalness as a matter of degree and look back at typical meaning postulates in that spirit, we find that meaning postulates are not *entirely* empty. Even the most vapid of them affords at least a bit of generality: "$(x)($**BACHELOR**(x) → ~**MARRIED**(x))" subsumes "If Reg is a bachelor Reg is unmarried," "If Irv is a bachelor Irv is unmarried" and "If Ronald Reagan is a bachelor Ronald Reagan is unmarried" as lexical truths.

I would contend that such humble subsumptions are authentically explanatory, though the explanations provided are very shallow. Why does "Ronald Reagan is a bachelor" entail "Ronald Reagan is unmarried"? Because the meanings of "bachelor" and "married" are such that "N is a bachelor" implies "N is married" no matter what singular term is uniformly substituted for "N." Why does "Ronald Reagan is unmarried and he believes in kirlian photography" entail "Ronald Reagan is unmarried"? Because the meaning of "and" is such that "S and T" implies "S" no matter what sentence is uniformly substituted for "S" and "T." There is no difference in kind between these two explanatory statements, but only a difference in degree of generality.

Quine thinks of meaning postulates as offering only "dormitive virtue" explanations. So do I. But against Quine and Moliere, I maintain that "dormitive virtue" explanations are genuine explanations, even though slight ones. When Moliere's physician proposes that laudanum puts people to sleep because it has a dormitive virtue, he succeeds in ruling out some alternatives, even though they are not very interesting or appealing alternatives. For laudanum to have a dormitive virtue is for any quantity of it to have or contain a power. People given laudanum have gone to sleep, not by chance, not by magic, and not even because a powerful god capriciously but uniformly chooses to put people to sleep when he notices that they have been given laudanum; they have gone to sleep because laudanum has something in it that *itself* puts people to sleep. Again, that is not an exciting or even very illuminating hypothesis, since few philosophers would have thought of doubting it, but it does rule out some conceivable alternatives and so is nonempty.[5] To explain is (at least) to rule out alternative generalizations, and to explain powerfully or illuminatingly is to rule out *more* and/or *whole ranges of* alternatives. The number of alternatives excluded and the number of ranges of alternatives excluded are matters of degree.[6] Thus meaning postulates and "dormitive virtue" explanations generally do genuinely explain; their shame is only that they do not explain very much. The Quine-Moliere objection, as originally stated, evaporates.

6. THE DAMAGE

But have we not done away with the principal motive, or at least Davidson's motive in "The Logical Form of Action Sentences," for doing truth-theoretic semantics in the first place? That motive was to "capture" felt implications *by assimilating them to "logical" implications*--more boldly, by showing them to *be* logical entailments, really, beneath the surface. Recall the wonderful rush of understanding produced by Davidson's analysis of action sentences (whether or not one agreed with that analysis' details, as I am sure no one did): The rendering of adverbial modifiers as logical conjuncts was a gorgeous paradigm case of reducing the unfamiliar (and mysterious) to the familiar by means of a clever hypothesis. Only the maddest logical conservative would think that Davidson's extensionalist treatment would work for all of a natural language--that the first-order predicate calculus alone is the underlying logic of English in particular; but it was not farfetched to imagine that first-order logic supplemented by some further logical operators would do. And if such an underlying logic were recoverable, then Davidson's great pattern of semantic explanation could have been extrapolated to all of human language.

But the Davidsonian project still presupposes the distinctness and the distinctiveness of a smallish identifiable class of genuinely logical operators. That presupposition of "genuinely logical" operators is just what I have been giving up. Moreover, it seems we must lose an important point of testability in a semantic theory: If there is no principled difference between a rule of logic and a mere "meaning postulate," and if any felt implication can be "captured" on the spot by the proclaiming of a meaning postulate, then felt implications are not the remorseless, implacable proving ground for semantics that they were intended to be.

Finally, our Cresswellian concession lets much of the air out of paraphrase arguments in semantics. It is usually supposed that if two sentences are synonymous or mutually paraphrase each other, then (not necessarily but) probably they share the same or logically equivalent logical forms. This assumption guides theorizing in a useful way. But again, if there is no principled difference between a rule of logic and a mere "meaning postulate," our two sentences may be related only by meaning postulate and have nothing at all in common that we would think of as formal. The idea of a "logical form" itself threatens to attenuate. (I say only "threatens," because it is kept from harm in Cresswell [1973] owing to that work's preservation of the nonEnglish intensional abstraction operator λ as a truly logical constant.)

7. DENOUEMENT

What is left? We have abandoned the idea of a small subclass of genuinely logical words within the English lexicon. There are only degrees of "logicalness," which I have proposed to understand as degrees of generality of application. There may be one or more significant psychological distinctions to be made along the continuum, but that cannot be known *a priori* and in any case is of no concern to the logician.

One can of course join Cresswell in positing a nonEnglish logical operator and tracing all implication in English to that operator. That is an attractive option, and preserves the idea of English as a formal language. Failing that, a semanticist is left, in regard to the capturing of implications, only with the aim of explaining implications in *as* "logical" terms as possible other things being equal. That is, the methodological instruction to explain implications logically and eschew "meaning postulates" gives way to a preference for the more logical, which on my view is just a special case of one's common theoretical preference for the more general and the more unified. That preference is no biting empirical desideratum, but is weak and very easily overridden.

Perhaps Gilbert Harman foresaw that such a weakening would be necessary when he expressed one of his early principles of theory preference as, "[M]inimize [nonlogical] axioms" (Harman [1972b, p. 42]). A straight-down-the-middle Quinean on the issue of meaning postulates would have said no such thing, but would simply have reminded the reader that so-called "nonlogical axioms" are bunk. On the other hand, the idea of *minimizing* nonlogical axioms presupposes that there are identifiable nonlogical axioms to be minimized, and unlike Harman we have abandoned that presupposition as well. There are neither logical operators with their mighty recursion clauses nor nonlogical axioms written in a special box, but only operators with their truth rules differing only in their generality of application. The relevant methodological instruction is simply to prefer the more general operators where one can, and as we have seen, that instruction has little authority.

Thus Clark's (1970) response to Davidson and his competing treatment of adverbial modification are considerably more attractive than some of his critics have granted. Clark botanized adverbial modifiers into a number of different categories and wrote a separate truth rule for each. Harman (1972b) and others saw those truth rules as nonlogical axioms, and faulted Clark for proliferating same. But though they are far from being "purely" logical, Clark's rules have a fair degree of generality, and are not just vapid bottom-level "meaning postulates" either; they are somewhere in the middle, and that is quite all right.

These findings have in no way dimmed my admiration for "The Logical Form of Action Sentences." But they have taken away my hope that Davidson's achievement can be replicated throughout all fragments of natural languages everywhere--unless the psychological "middle ground" hypothesis formulated at the end of section 3 above should turn out to be true, as is unlikely.

NOTES

[1] Whose roots are in Plato, Leibniz, Frege, Russell, Wittgenstein's *Tractatus*, and the early writings of Hintikka and Montague-- Davidson's own distinctive contributions having been (i) to implement all those theorists" notions of correspondence and/or picturing by way of Tarski's theory of truth, and (ii) by (i) to make semantics into a testable enterprise.

[2] "Unlike any of the symbols we have met so far, λ is a logical constant with a fixed interpretation" (p. 84). The principles of λ-conversion are "the only principles which are independent of a particular value assignment" (Cresswell [1974, p. 470]).

[3] I am less sympathetic to Cresswell's succeeding example and argument, to wit:

If we try to mark off a class of entailments which depend only on the "logical words" of English we are faced with the invidious task of deciding what these are. E.g., is *if* one of them? If it is and if, as seems likely, *if* in English is not truth-functional then our logic will have to be some, as yet undiscovered, highly intensional logic. It is of no use to reply here that *if when understood as material implication* is a logical word of English unless we can sort out those sentences in which it is being so understood....

I do not see the force of this. Why does non-truth-functionality automatically make an operator a problem case, unless one simply means to stipulate (as Cresswell expressly does not) that only truth-functional operators count as logical? Nor does the undiscoveredness of the intensional logic militate against "if"'s being genuinely logical, unless *having been discovered* is stipulatively and rashly required.

[4] Harman (1972a) appeals to this criterion in arguing that modal operators are not logical constants.

[5] For a similar defense of dormitive virtues against Moliere, see Sober (1982).

[6] Though here as ubiquitously in philosophy, the idea of degree along with that of proportion is officially stymied by the nondenumerability of the containing population under a suitably fine-grained mode of individuation.

CHAPTER 11

PROPOSITIONS AND ANALYTICITY

It has been nearly forty years since the publication of "Two Dogmas of Empiricism" (Quine [1951/1961]). Despite some vigorous rebuttals during that interval,[1] Quine's rejection of analyticity still prevails--in that philosophers *en masse* have either joined Quine in repudiating the "analytic"/"synthetic" distinction or remained (however mutinously) silent and made no claims of analyticity.

This comprehensive capitulation is somewhat surprising, in light of the radical nature of Quine's views on linguistic meaning generally. In particular, I doubt that many philosophers accept his doctrine of the indeterminacy of translation, which directly underwrites the rejection of analyticity even though it did not figure prominently in "Two Dogmas" but was made explicit only in *Word and Object* (Quine [1960]) and subsequent papers.[2] (Indeterminacy of translation spells the death of individual word meaning as well as that of propositional or locutionary meaning, and so leaves no room for *truth by virtue of word meaning* as is required by analyticity.)

In this chapter and the next I shall make a Quinean case against analyticity, without relying on the indeterminacy doctrine. For I would like to join the majority in denying both analyticity and indeterminacy: *Contra* Quine, I think there is determinate propositional meaning, as represented by a structured set of possible worlds, but with Quine I doubt that any sentence of a natural language is true solely in virtue of its meaning.[3] However, we shall see that the attack on analyticity is partly blunted by the actual practice of stipulative definition, and we shall also have to concede a sense in which some sentences are "true in virtue of meaning," though it is not a philosophically important sense; those issues will be the focus of Chapter 11.

1. QUINE AGAINST ANALYTICITY

As is well known, Quine begins by complaining that the "analytic"/"synthetic" distinction is imprecisely drawn. It should by now be equally well known that that is the least of Quine's complaints. He further objects to the circularities that infest philosophers' informal explanations of the "family circle" of intentional notions he is attacking: "analyticity," "synonymy," "contradiction in terms," and so on. Grice and Strawson (1956), among many other commentators, objected that *all* systems of definitions have this "circular" character. But Quine's point is that ordinary systems of definitions normally contain at least one term which is antecedently and nontechnically understood, in which case the circularity (of course) does not matter, while among the family-circle words there is no such term--they are all technical. (One might think that "synonymy" is nontechnically clear enough, but as Harman (1967, pp. 135-137) has emphasized, ordinary people use that word far more broadly and loosely than philosophers do.[4] One might also think that "definition" is all right, but as we shall see, "definition" is a morass.)

Yet the real objection to analyticity is neither the imprecision nor the unspecifiability of a distinction between analytic and synthetic; the first half of "Two Dogmas" was historically very misleading in this regard.[5] *Defining* analyticity is not the problem. In fact, analyticity is rather easily defined, as a strongly modalized variety of truth: not just nomologically or even metaphysically necessary truth, but "conceptual" truth, truth by virtue of lexical meaning alone. A sentence is analytic iff its own meaning suffices to make it true, regardless of any (other) contribution from the world.

This way of understanding analyticity has obvious epistemological consequences: An analytic sentence would be *unrevisable*, in the sense that to deny or reject it would be *eo ipso* to abandon its standard meaning; one who called it false would be, as Quine says, not denying the doctrine but changing the subject. Thus nothing could count as evidence against the truth expressed by an analytic sentence, and more generally we could have no rational grounds for doubting that truth (we could be mistaken only about the meanings of the relevant words).

Quine's real complaint is that analyticity thus understood is just unexemplified. There are no analytic sentences, because no truth is necessary in any stronger sense than the nomological, and no belief is unrevisable. (There is, if you like, "analyticity" relative to a theory on an axiomatization of that theory: A sentence may be *treated as* analytic within a theory, by being used as an axiom and introduced in textbooks by an axiomatic definition. But mere analyticity-within-a-theory-on-an-axiomatization is not what partisans of analyticity have had in mind.)

So let us review some attempts to explain how a sentence can be both necessary in some stronger sense than that of physical or causal necessity ("logically," "conceptually," etc.), and epistemically unrevisable in the way described. There are three main kinds of account expressed or presupposed in the literature. The first two I shall mention but dismiss briskly. It is the third that will detain us.

2. THE FIRST UNPROMISING ACCOUNT

The first account is a straight shot: the idea that truth in virtue of meaning is just that. Sentences have meanings; some meanings are such as to guarantee truth, and that is (virtually) all there is to it. Call this the "Simple Account."

At the time of "Two Dogmas" *meanings* were thought of as propositions, language-neutral abstract entities "expressed" by sentences. The positing of propositions was supposed to explain the properties of language-bound sentences individually and in pairs--meaningfulness, ambiguity, synonymy, entailment, and the like--in terms of the relation of expressing. Psychological facts about the propositional attitudes were complementarily to be explained with the aid of the further relation of "grasping" between a subject and a proposition.

Understood in these terms, the Simple Account of analyticity has it that a sentence is analytic just in virtue of expressing a proposition of a special sort--the null proposition that rules out no contingent possibility. Alternatively, one might say that although an analytic sentence is well-formed and trivially or degenerately true, it expresses no genuine proposition. Either way, such a sentence--meaning what it does mean--cannot be false.

Quine's objection to this propositional theory of analyticity is that it invokes propositions. His main mature objection to positing propositions is that if there really were propositions then translation would be determinate; yet translation is indeterminate. However, I myself am sworn not to rely on the indeterminacy doctrine, and so must find other objections. Quine himself makes two, and a third can be pressed on his behalf.

The first objection is akin to the indeterminacy claim, but does not rely on it: The proposition theory does not square with the translation habits of ordinary people and with their ordinary talk of synonymy and the rest. In the real world, people count as "synonymous" any two sentences that serve roughly the same contextual purpose on an occasion, and such pairs will seldom be candidates for doing what philosophers call expressing the same proposition. Ordinary people rather engage in the loose free-for-all practice that Quine calls "paraphrase" ([1960, sec. 33, p. 208, and secs. 53-54], and see Harman, p. 142). Secondly, propositions themselves are *entia non grata*, "disorderly elements," "creatures of darkness," etc.; they are nonspatiotemporal and acausal, their identity conditions are inscrutable, and so forth (Quine [1960, Ch. Six; 1969c]).

The third objection is emphasized on Quine's behalf by Harman. It is that propositions are *flatus vocis* or dormitive virtues, and do no explanatory work whatever. ("Why are these two sentences synonymous?" -- "Because they express the same proposition." -- "Oh, I see.") The proposition "theory" of meaning is a pseudo-theory, not a substantive theory at all.

I disagree with Harman on this general point, for I think the theory of propositions is a genuine and even substantive theory, even though a very bad one (Lycan [1974]). But there is a serious question of the explanatoriness of the proposition theory with respect to analyticity in particular. A sentence is supposed to be true, and true in virtue of meaning, because it "expresses the Null Proposition." Like Harman, I fail to discern the explanatory force of that "because."

Even if we know what "expressing" is, what do we know of the "Null Proposition" except that sentences that express it are supposed to be analytically or conceptually-necessarily true?

One may of course attempt to answer that rhetorical question nonrhetorically, by offering a story about what "propositions" are that says in particular what makes the "Null Proposition" special. The possible-worlds story comes instantly to mind, for propositions can be taken to be sets of worlds. Indeed, David Lewis' theory of analyticity is an instance of the Simple Account by way of propositions by way of possible worlds: For Lewis, a sentence is analytic iff it is true at every world--and so expresses, though Lewis does not use the term, the Null Proposition, i.e., the proposition that excludes nothing.[6] Thus Lewis avoids my foregoing criticism of the Simple Account, and advances the issue by explicating analyticity as truth in all possible worlds. An analytic sentence is true in virtue of expressing the proposition that it does, the Null Proposition, for that proposition is the set of *all* worlds and is nowhere false.

Lewis maintains that the truth of a proposition throughout all worlds is a mind-independent fact of the pluriverse. Thus for him an analytic statement is true in virtue of the modal structure of the world, in virtue of what is objectively possible. In a sense, then, although a Lewis-analytic sentence is true in virtue of its meaning, it is not true *solely* in virtue of that but partly in virtue of an objective feature of the pluriverse.[7] That is perhaps only a quibble, but in any case, there ensues the more trenchant Quinean question: Why should anyone think there are any sentences that are true in all possible worlds, if "possible" means something seriously weaker than nomologically possible? Lewis' theory makes no headway at all against this most basic misgiving.[8]

Might one perhaps pursue the Simple Account *without* reifying meanings, i.e., without appeal to propositions at all? That would be to hold that the truth of certain "necessary" sentences is still explained *solely* by reference to considerations of meaning, though now in some ontologically non-toxic sense. But if the Simple Account is to be left at that, we are given no hint of how the explanation of truth by meaning alone could proceed. We are told "how" sentences can be true by virtue of their meanings alone, only by being told that those special sentences' meanings suffice to make them true. Unless some entities comprising a mechanism are posited--at least propositions and/or possible worlds--no explanation emerges at all.

3. THE SECOND UNPROMISING ACCOUNT

The second unpromising account of analyticity--which historically replaced the Simple Account so far as the latter modestly appeared in Russell and Moore--is the positivist/verificationist theory of truth in virtue of *unfalsifiability*. A sentence that is verified by every possible experience, or at least falsified by no possible experience, would be a candidate for analyticity. And this Verificationist Account avoids Quine's and Harman's objections to the Simple Account, for it is genuinely substantive and explanatory. Verificationism in general is an undeniably explanatory theory of meaning, even if a false one, and it entails in particular that if a sentence is radically unfalsifiable, there is something very special about that sentence's meaning: so far as the sentence is true, it is true solely in virtue of meaning (for its meaning, i.e., its verification-condition, affords no room for falsity).

There are basically three objections to the Verificationist Account of analyticity. First, how could we ever know that a sentence had this character of in-principle unfalsifiability? At least, how could we possibly know that a sentence was immune in this way to refutation, *unless* we had some deeper antecedent reason to think the sentence was impregnable, e.g., that it was already true by convention. A look at the more accessible Positivist writings suggests that irrefutability is not itself the *explanation* of a truth's "conceptual necessity"; rather it is explained *by* necessity, which is in turn explained by something more familiar.

It is sometimes urged against such points that we know a sentence to be unfalsifiable simply because we cannot imagine or conceive the sentence's denial. Quine's answer to someone who pleads just plain inability to imagine a thesis' being false is simply that the person has a poor imagination. What one can imagine at any given time is almost entirely dependent on one's current theory, as the history of science and the history of ideas indicate; even the great intellectuals of the past were "unable to imagine" many of the things which today are commonplaces.[9]

In any case (second objection), the operative term in the slogans "verified by all possible experiences," "falsified by no possible experiences" and the like is, of course, "*possible*." The Verificationist Account cannot do without this qualification, and "possible" is a member of the family circle. Explaining necessary truth in terms of "possible experiences" is much like explaining it in terms of "possible worlds." Once again we must ask why certain imagined experiences-- e.g., as of seeing an object that is simultaneously red all over and green all over, or as of a square circle--are supposed to be impossible in any stronger sense than that of nomological impossibility. To insist that such experiences just are inconceivable in that very strong sense adds nothing to one's initial insistence that the corresponding negative sentences are analytic.

Quine's most famous objection to the Verificationist Account (and my third) is a corollary of his Duhemian point against atomistic verificationism generally: that "statements about the external world face the tribunal of sense experience not individually but as a corporate body" ("Two Dogmas," p. 41). Any single sentence of our total theory may be retained in the face of apparently recalcitrant experience, if enough more or less drastic revisions are made elsewhere in the theory. Thus no individual sentence *has* its own specific verification-condition. Someone might reply that while Duhem's point is true of contingent and empirical sentences, it obviously fails for sentences which are true or false no matter what; but that would, as before, beg the question of whether there are any such sentences in the first place.

Happily (for them), the Logical Positivists had a positive account of analyticity, in addition to experiential unfalsifiability, their negative criterion. It was that analytic sentences are true *by convention*. The Convention Account of analyticity is the third and last view that I shall consider. Notice that it is entirely independent of the Verificationist Account, even though the Positivists held both;

historically, it was preserved among the Ordinary Language philosophers who rejected verificationism, and continues to be accepted by at least one (otherwise) hard-core scientific realist, Armstrong (1981, especially Chs. 5 and 6).

A self-respecting contemporary metaphysical realist and referential semanticist finds the idea of "truth by convention" grotesque. How could a sentence be *true*, i.e., correspond to reality in a referential manner, entirely *by convention*? But of course the Positivists and the Ordinary Language philosophers were not (thoroughgoingly) either metaphysical realists or referential semanticists. More to the point, the idea of truth by convention is defensible and the Convention Account of analyticity is far more compelling than was either of its two predecessors.

4. TRUTH BY CONVENTION

There are three main ideas of truth by convention. One is that of lexical meaning as codified in authoritative dictionaries of natural languages. The second is that of lexical rules or "meaning postulates" codified in formal languages. The third is that of truth by *stipulative* definition. In the rest of this chapter I shall discuss the first two; the third will be the topic of Chapter 12.

The dictionary idea is that word meanings are conventional and that the governing conventions are codified by the entries in a good dictionary for the language in question. Any "information" contained in or easily read off from a proper dictionary is thus purely conventional, and cannot be doubted or questioned by anyone who has mastered the language.

An obvious objection to this idea is that, in the real world, dictionaries are not stone tablets bearing our conventions in axiomatic form. *Most* entries in a typical dictionary do not even purport to codify analytic truths, and none does anything whatever to distinguish mere factual information from what philosophers would consider strictly parts of the meanings of the *definienda*; usually the entries contain just informal explanations of how to apply the concept or recognize the item in question, recognition criteria mixed with a motley

PROPOSITIONS AND ANALYTICITY 257

of commonsensical facts. Hence Quine's apt remark that dictionaries differ in no principled way from encyclopedias.

One might respond that in real life, lexicographers get sloppy (many of them are free-lancing, and all are underpaid); that is just real life for you. But Quine has a more powerful objection to the dictionary idea: Considering the suggestion that ""bachelor'...is *defined* as "unmarried man'," Quine asks, "Who defined it thus, and when?" ("Two Dogmas," p. 24) This question is sometimes passed by as a rhetorical flourish or as merely a superficial piece of sarcasm, but it has two deep points. One is that on pain of regress, conventions of language could not all have been established explicitly by syndics sitting in council. (There are infinitely many logical truths. But there could not be a complete infinitely totality of single conventions which fix all the logical truths, one by one, immediately. Therefore if logical truths are true by convention, they must be given by general conventions, we must rely on an antecedent knowledge of the very same sort of logical truths, in mediating our inferences. Thus the regress.) Quine originally took this to embarrass the idea of a linguistic convention quite badly, but he agrees it has been circumvented by Lewis' (1969) account of tacit convention.[10]

The second and more telling point is that the lexicographer is in exactly the position of the field linguist in a case of radical translation; the lexicographer merely observes language already in use and theorizes about it. Thus he or she cannot experimentally distinguish between sentences that are fixed by convention as "analytic" in the target language and those that are merely considered *obvious* by the native speakers, since all that meets the eye and ear is one class of "stimulus-analytic" strings, the sentences that command universal assent and whose hypothetical denials elicit funny looks.

Granted, a lexicographer's evidence base is not exhausted by the linguistic behavior of others. The lexicographer who is a native speaker may appeal to his or her own linguistic competence and internal sense of word meanings. But the same criticism applies to that appeal, as it does to the Ordinary Language philosopher's introspective test of meaning ("Would we still call a thing an X if it weren't a Y any more?"): We cannot say even of ourselves, simply by introspecting, whether some belief of ours is true by definition or whether it is merely a very well-entrenched truism. For we have nothing to go on but the feeling of obviousness and thorough-entrenchedness. One might say that if the lexicographer knows what genuine synonymy is, he or she is entitled to use that notion in constructing an ideal, nonsloppy dictionary of a language, and can make the appropriate distinctions on sight. But this notion of "genuine synonymy" is just what is in question, and claimed by Quine to be an unexemplified philosophers' artifact. The appeal to existing or even idealized real-world dictionaries cannot discredit Quine's claim.

Perhaps ordinary natural language is not the proper locus of analyticity. Formal languages are better suited to foment analyticities, since they are under the control of logicians and philosophers in a way that real natural languages are not. And Carnapian "semantical rules" or "meaning postulates" are splendid candidates for analyticity. They are assimilated to the *definitional postulates* of an axiomatized theory.[11] A logical consequence of a definitional postulate, or at least the postulate itself, ought to be true in virtue of meaning alone.

Quine has a good deal to say about this;[12] two criticisms emerge. First, postulates are postulates only relative to a given context of inquiry; they can be changed around at will without this effecting any substantive change in the theory they inhabit. Notoriously (Hanson's [1965] example), the second law of classical mechanics, $F = m(d^2s/dt^2)$, can be taken or employed variously in the organization of theory--as an axiom, as a stipulative definition of "F," as a definition of "m," or as a purely empirical generalization, depending merely on how the containing theory is axiomatized.

A Positivist might well reply here that postulates can indeed be added and dropped at will for reasons of convenience, and theories re-axiomatized, but that such tampering necessitates corresponding changes in the meanings of the terms which occur crucially in (and are "implicitly defined by") the theories; different axiomatizations give different meanings to the theoretical terms, even though the theories do not vary in overall empirical content. But Quine rejoins that there is no independent evidence at all of any such meaning changes' having occurred, when the postulates of actual theories have been readjusted in real life. Put the matter up to any scientist or any intelligent but philosophically untainted specialist in some other cognitive discipline, and neither informant will know what we are talking about; though they might grant loosely that "what we understand by" a theoretical term has changed when we reaxiomatize or introduce a new definitional postulate, they would never maintain that the term had "just plain changed its meaning," or that it now simply meant something different from what it meant before. (Hilary Putnam gives numerous examples of relevant scientific changes that plainly have involved no meaning changes in any but a question-begging sense.[13]) The Positivist insists that the changes occur, because they must occur if the Positivist theory is correct. But on pain of severe *ad hoc*ness, the meaning changes would have to be exhibited in a non-question-begging way.[14]

Quine's second criticism of the appeal to definitional postulates is that theories can turn out to be false; so "truth" by convention" does not guarantee *truth*. (Harman puts the criticism in this way on pp. 130-131.) As always, someone may reply that a theory's definitional postulates are special, in that even when their containing theory goes down in flames, their analyticity exempts them from falsification. But as always, the alleged analytic exemption is just what is at issue.

One might think that at least the theorems of classical logic would be exempt, before any special vocabulary had been introduced. And the notion that they hold purely in virtue of the meanings of the logical constants is a very appealing philosophy of logic, whether or not one approves of "truth by definition" in science or in other empirical discourse about the material world. But the present point holds for logical truths also: As the Law of Excluded Middle is questioned both (for epistemological reasons) by intuitionists and (for scientific reasons) by quantum logicians, so *might* we find epistemological or scientific reasons for qualifying even the Law of Noncontradiction. Not even logic is epistemically unrevisable in the strong sense required for analyticity. Moreover and contributorily, as we saw in Chapter 10, the alleged distinction of kind within linguistic semantics between the "logical constants" and other words of English is dissolved by even a little examination.

The third idea of truth by convention, based on stipulative definition, will require a new chapter.

NOTES

[1] Bergmann (1953); Grice and Strawson (1956); Bennett (1959); Katz (1967, 1974), and elsewhere; and of course many others.

[2] Quine (1968, 1969, 1969, 1970, 1987).

[3] Callaway (1980, 1984) has also championed meaning without analyticity.

[4] Harman's splendid essay (hereafter just "Harman") is the classic exposition of Quine's views on meaning; my own understanding of analyticity derives very largely from it.

[5] See Quine (1960, p. 65, n 3). I conjecture that all the charges of circularity and emptiness of characterization were really meant to defame the idea that behind philosophers' "intuitions" of analyticity lies a substantive and good theory of meaning that supports the intuitions and has explanatory merit of some sort. I agree with Quine that that idea is false, even though unlike him I believe there is a substantive and good theory of meaning.

[6] Lewis (1969, pp. 174-177)--a book on linguistic convention addressed in part to the analyticity question, in which, ironically, Lewis repudiates the understanding of analyticity as truth *by convention*. Though what proposition an English sentence expresses is (obviously) a conventional matter, Lewis insists that the truth at every world of the proposition expressed is not conventional at all but is a mind- and cognition-independent modal fact about the universe (p. 207).

[7] Against this, a referee has argued that since convention establishes what proposition a given sentence expresses, then if, for a particular sentence, convention establishes that that sentence expresses the necessary proposition, then convention *a fortiori* establishes its truth and so it is after all true by convention. But, I reply, it is not true (solely) by convention that that proposition expressed by the sentence, the necessary proposition, *is* the necessary proposition. So convention alone still does not guarantee the original sentence's truth.

[8] N.b., Lewis does not himself claim that there are any analytic sentences. Also, a response can be hazarded on his behalf. Suppose (waiving cardinality problems) that there is a set of all possible worlds. Any set of worlds is a proposition, and so the set of all worlds is a proposition, the Null or necessary one. But a proposition corresponding to a neatly delineated set of worlds is surely expressible in English, even if there are some sets of worlds that are just too heterogeneous to match single English sentences (which I doubt). Thus, some English sentences express the Null Proposition. Which ones? The standard philosophical examples of trivial verbal truths are the obvious candidates; therefore probably they are analytic. My objections to this argument are (i) that it does nothing to show *what it is* about the trivial verbal truths that makes them express the Null Proposition, and (ii) that in the face of Quinean skepticism we have no reason to think that there are multiple possible worlds distinguished from each other only by "logical" or "conceptual" possibilities that are not nomological possibilities.

[9] Nerlich (1991) points out that Leibnizian verificationist-*cum*-imaginability arguments against absolute space simply though tacitly assume geometrical theses that hold only for Euclidean space--e.g., that a given figure has similar figures of any size!

[10] The original objection was stated most fully on pp. 85-98 of Quine (1966c); the recantation occurred in Quine's Foreword to Lewis (1969).

[11] So far as natural languages can themselves be formalized, a Carnapian structure can be imposed; a formal "semantics for" a natural language such as English might contain a box of "meaning postulates" or "nonlogical axioms" intended to exhibit facts of lexical meaning. We have seen in the previous chapter that Carnap's practice is a vexed one.

[12] "Two Dogmas," pp. 35-36; Quine (1963, p. 392); and elsewhere.

[13] See Putnam (1975a, pp. 310-315; 1975b; 1975d; 1975e, pp. 254-257), and other papers collected in Putnam (1975c).

[14] I think, incidentally, that this rejoinder is what Quine has in mind when he makes a somewhat misleading remark on p. 37 of "Two Dogmas," calling the analytic/synthetic distinction a "metaphysical

article of faith." He means, not that the distinction is a prediction that may or may not come true, but that it is something that is dogmatically believed come what may in the absence of any convincing evidence at all. Pp. 122-127 of Quine (1963) bear a similar interpretation.

CHAPTER 12

STIPULATIVE DEFINITION AND LOGICAL TRUTH

Let us turn to our third idea of truth by convention, inspired by the occasional practice of *novel* stipulative definition within more or less natural languages.

1. STIPULATIVE DEFINITION

Suppose I offer a novel coinage:

Dudent $=_{df}$ Dumb student.

And suppose I add, now employing my newly augmented object language, "All dudents are students, you see." How could I possibly be mistaken? My assertion seems a clear example of the truly unrevisable. To question it would be to question either the success of the wisdom of my stipulation, and to reject it would be to reject the stipulation and refuse to join in speaking the augmented language; my audience could not "deny the doctrine" and genuinely disagree with what I had said, for the ostensible denial, "There are dudents who are not students," simply makes no sense in the context. It is either a straightforward self-contradiction or it means nothing at all.

Quine himself is startlingly respectful of this idea:

There does...remain still an extreme sort of definition which does not hark back to prior synonymies at all: namely, the explicitly conventional introduction of novel notations for purposes of sheer abbreviation. Here the definiendum becomes synonymous with the definiens simply because it has been created expressly for the purpose of being synonymous with the definiens. Here we have a really transparent case of synonymy created by definition; would that all species of synonymy were as intelligible. For the rest, definition rests on synonymy rather than explaining it. ("Two Dogmas," pp. 25-26)

This apparent concession is larger and more damaging than perhaps Quine realized in making it. For it seems to allow that there are analyticity-generating synonymies and so genuine instances of full-fledged analyticity, even if they are few. Moreover and worse, if the introduction, for notational convenience, of new abbreviations for familiar but cumbersome terms can yield analytic truths, then the way is opened to a whole further range of analyticities. For if we are ever able to make two sentences equivalent in meaning by fiat, then presumably it is also possible for us to find out conclusively, to hypothesize, or even to "reconstruct" *à la* Hobbes and the State of Nature, that someone already has made other pairs of sentences equivalent by fiat in the past. So, it seems to me, a good Quinean ought to argue as powerfully as possible against the possibility of creating genuine analyticities even by explicit stipulation. Accordingly I shall take up and try to expand some of Quine's and Harman's objections to the Stipulative Definition account of analyticity, adding two of my own as well.

2. THE OBJECTIONS

The obvious Quinean objection is based on the indeterminacy doctrine: Suppose a friend comes by and offers a stipulative definition, say: "'Veline' shall mean *vegetarian cat*." We could accept a translation manual according to which our friend's phrase "shall mean" is taken onto the term "obviously denotes" and suitable readjustments of syntax are made. The idea is that our friend, in her role as unspoiled native, behaves toward the indicative in question, "All velines are vegetarian cats," exactly the same as she behaves toward an admittedly synthetic but trivial and obvious truth--except for the attachment of a particular label: "analytic," " $=_{df}$," "shall mean," or what have you. And our translation of the *label* is indeterminate; there is no fact of the matter. And so there is no fact of the matter as to

STIPULATIVE DEFINITION 265

whether our friend's utterance is really a stipulative definition or merely the reiteration of an obvious truth for the common good.

On Quine's view, her utterance is indeterminate in a second way as well: There is no fact of the matter as to what her putative *definiens* means. Even if we were able to bypass the indeterminacy of "$=_{df}$" & Co., and were able to say determinately that "veline" was analytically connected to another expression, we would not thereby know a necessary truth unless we knew what the *definiens meant*, to the exclusion of what it coextended with but did not mean--and this, according to the indeterminacy doctrine, we cannot know, because there is no fact of the matter to be right or wrong about.

As before, I shall not rely on the indeterminacy doctrine. But the first of these two appeals to translation has some force independently of the indeterminacy claim.

Objection 1: On p. 27 of "Two Dogmas," Quine lists three possibilities as to what is going on when new notation is stipulatively defined in a formal system. Of the first two (both technical equatings of long technical locutions with shorter terms, but for different purposes) he complains, as usual, that they already presuppose a prior notion of strict, analyticity-generating synonymy and so cannot be appealed to in an argument for the existence of analytic sentences.[1] In keeping with his concession quoted above, he expressly refrains from bringing the same complaint against the third possibility that "...the definiendum may be a newly created notation, newly endowed with meaning here and now."

But we may bring it. The idea of adventitious stipulative definition *as it would have to be construed for the purpose of saving analyticity* must presuppose the notion of strict, analyticity-generating synonymy. Otherwise it would collapse back into mere definitional postulation. Stipulators must have the power to create synonymies and truths by fiat. The "dudent" example seems to show that people do sometimes have that power. But notice that the "dudent" argument still presupposes the analyticity of logical truths: "All dudents are students" is true because "dudent" has "student" as part of its meaning and all *students* are students (cf. Quine [1966c], pp. 71-72). As we have seen,

CHAPTER TWELVE

Quine has argued against the analyticity (in the strong sense) of the logical truths; so he would hardly give in on that of sentences obtained from logical truths by substitution. Recall Harman's point that a sentence's being "endowed with meaning" on the spot does not prevent it from being false and thus does not *solely* explain its being true.

Notice too that if "All dudents are students" is truly analytic, its syntactic denial is meaningless; if a sentence really is true *just* in virtue of its meaning, then that sentence's syntactic denial cannot mean what it appears to say, and so either means something else or means nothing. But "There are dudents who are not students," though (if you like) logically false, is not meaningless. We all know what it means; if we did not, we would not know that it was logically false and that to token it is to contradict oneself.

Objection 2: Let us revisit and generalize the first of Quine's two arguments against analyticity through definitional postulation (cf. Harman, p. 141). The main problem was that the same theory or belief system can be axiomatized in different ways, treating different elements as "analytic." More formally: Suppose

(1) A sentence S in theory T_1 is "true by definition."

And

(2) The very same sentence S appears in theory T_2 as a purely empirical generalization.

(Let S, for example, be "$F = m(d^2s/dt^2)$.")

(3) T_1 and T_2 are merely two different axiomatizations of more or less the same body of truths.

(4) A sentence's analyticity/syntheticity is a function of the meanings of its component terms. [Tenet of the "analyticity" view].

STIPULATIVE DEFINITION 267

(5) Analytic sentences are those which are "true by definition"; others are synthetic. [Assumption for *reductio*]

But

(6) The meanings of S's component terms do not change (in any ordinary sense) from T_1 to T_2. [Seemingly supported by (3), but also plausible on its own]

(7) Either S is analytic in both T_1 and T_2 or S is synthetic in both T_1 and T_2. [(4),(6)]

(8)' S is analytic in T_1 but synthetic in T_2. [(1),(2),(5)]

(9) CONTRADICTION! [(7),(8)]

(10) (5) is false. [Since our premises (1), (2), (3) and (6) are true and we grant the analyticity theorist (4)]

So "definition does not hold the key to analyticity."

This formulation of the Quine-Harman argument opens the way for the standard reply simply put: that premise (6) begs the question, because on the "analyticity" view it follows trivially from (1) and (2) that the meanings of the component terms of S do change.

Here as in the previous, more specific case of definitional postulation, I am in sympathy with Quine's and Harman's replies on this point: Who says the meanings change from T_1 to T_2? It *seems* that all we have here is two organizations of the same theory differing only in elegance or in convenience for particular purposes. It is up to the defender of analyticity to show why that natural reaction is wrong. But the analyticity theorist cannot do this without relying on the essential tenets of just the philosophical view that is in question. Thus, it is the analyicity theorist, not Quine and Harman, who is begging the question here if anyone is.

CHAPTER TWELVE

Just as there is no difference between a "meaning postulate" and an ordinary postulate (in the sense of a very general truth chosen to generate other truths), that has other than *ad hoc* considerations to recommend it, there is no non-*ad-hoc* difference between a sentence "true by stipulative definition" and one simply true *ex hypothesi*. Even stipulative definitions, like postulates, can rationally be given up in physics without apparent change of meaning. Although the clearest cases of postulational change without meaning change involve natural-kind terms when new empirical information comes in, there are other, more ordinary occasions for semantically harmless readjustment of taxonomies as well.[2]

Objection 3: In stipulative definition, there is an unexpected problem of *getting from* the stipulator's token to an indicative sentence held to be analytically true. The problem takes a bit of explaining.

One strong temptation to see stipulative definition as yielding (however harmless) analytic truths arises from the fact that, when a person utters (tokens) an explicit stipulative definition, one cannot at all appropriately respond, "That's false." (Lucy: "Here's my first definition: A 'freebish' is a dog eating pizza." Linus: "That's false, Lucy, because..." Sound effect: *POW!*) Naturally enough, this is felt to show that explicit stipulative definitions and their consequences *cannot be* false, and thus are necessarily true.

I shall show here (expanding on pp. 71-72 of Quine [1966c]) why this approach to necessary truth is blocked. Let us first look at some typical instances of explicit stipulative definition:

(a)   "Veline" $=_{df}$ vegetarian cat.

(b)   All velines are by definition vegetarian cats.

(c)   We shall use "veline" to mean "vegetarian cat."

(d)   Let "veline" abbreviate "vegetarian cat."

(e)   "Veline" shall denote vegetarian cats.

(f)   Let "veline" mean *vegetarian cat*.

Now none of these instances is *itself* a straightforward indicative sentence alleged to be analytic. It is the prior occurrence of one of these that it supposed to legitimize the imputation of analyticity to some later, distinct sentence: "All velines are vegetarian cats." So we may well ask how this process takes place, and what sort of relation holds between the stipulative sentences (let us hereafter call them just "stipulatives") and the indicative, "All velines are vegetarian cats."

Stipulatives (d) and (f) best capture the spirit of convention-fixing, being hortatory subjunctives. Obviously they cannot be true or false; they state *rules*, more or less in the form of *commands* from the Great Convention-Giver to her people. (c) and (e), I think, are best interpreted as announcements of the speaker's intention to obey a certain rewriting rule--the idea being that if you want to understand the speaker, you must obey the same rule *pro tem*. (c), then, is just a slightly more tactful version of (d).

It is hard to assess (b)'s logical status; (b) seems to be an indicative, not a subjunctive or a command. But one could, I suppose, state the stipulative definitions of one's theory in this form, so long as one used a heading on the page which made it clear that the sentences codified stipulative and not reportive definitions. (a) could be read aloud as almost any of the other stipulatives listed; it does not seem to have a separate logical status of its own.

So we have two basic kinds of stipulative: A hortatory subjunctive, or perhaps an imperative telling us to do such-and-such under such-and-such conditions; and an indicative sentence that either is flagged (by a key phrase such as "by definition" appearing within it) or is prefaced by a heading such as "Abbreviations" or "Defined Terms" or even "Semantical Rules." In what way do stipulatives of either kind yield necessary truths?

Let us take the latter kind first. (b) does seem to entail "All velines are vegetarian cats," but what entitles us to conclude that the latter sentence is necessary in the sense of analyticity? Only the presence of the flag, the phrase "by definition," or perhaps the appearance of (b) under a heading of the kind just mentioned. And it is the forces of just such phrases and headings that is in question here, the question of whether a speaker can make a sentence true by announcing that it shall be so. So the stipulative occurrence of (b) does not suffice to distinguish (b) from the other stipulatives or to show that "All velines are vegetarian cats" could not possibly be false. So let us go on to the more promising possibility of getting analytic truths directly from (rewriting) *rules* of the language itself.

How is "All velines are vegetarian cats" obtained from, say, (d), or from a still more explicit rule, "Rewrite 'vegetarian cat' as 'veline'?" There is no entailment here (even if we could make sense of the notion of entailment independently of all the talk about meaning and necessary truth); since (d) is not an indicative, it has in the ordinary sense no truth-value and hence cannot entail anything. The only obvious way to get "All velines are vegetarian cats" from our rewriting rule is to regard it as being the result of applying the rule to the logical truth, "All vegetarian cats are vegetarian cats." But there is still no reason to regard "All velines are vegetarian cats" as being true by convention, even though it is in part the product of a rewriting rule or conventional abbreviation which *is* (we may concede) in a clear sense purely a linguistic convention; because it has not been shown that the logical truth originally operated on by the abbreviative rule is true by convention. We could regard "All velines are industrious" as being the result of applying our rewriting rule to "All vegetarian cats are industrious," but that does not make the former a necessary truth.

A second way of putting Objection 3 is to point out that any way of getting a truth couched in the object language from an explicit or tacit stipulative must make a significant use/mention move and that that move is sure to be fallacious. Stipulatives are metalinguistic, besides being hortatory or proclamatory in mood, while (in terms of truth conditions) the corresponding object-language trivialities are directly about the world, however little they illuminate it. Therefore the object-language truisms do not follow deductively from the stipulatives; and there is no *other* reason to think that the stipulatives make them analytic. (There is no objection to calling them "definitional truths," so long as it is understood that "definitional" does not mean "analytic." Nor need we balk at calling *expressly* logical

STIPULATIVE DEFINITION 271

truths logical, any more than Quine would object to calling laws of physics nomological truths, or to calling laws of the state of Massachusetts legal truths. The theorems of a particular system of logic form a well-defined and interesting class. It is just that they are not true (solely) by definition or by virtue of meaning.)

Incidentally, we now have a way of blocking the argument from the inappropriateness of "That's false," mentioned above: The reason "That's false," uttered in response to a stipulative definition, is inappropriate or sounds funny or whatever is not that stipulative definitions cannot-be-false in the sense of analyticity. They are barred from falsity in the trivial sense of being truth-valueless by virtue of their syntax. In the same sense they cannot-be-true.

I close my Quinean case with an objection to "truth by convention" generally: It is only recently that anyone--Lewis[3]--has made clear sense of the notion of a "convention" in the first place. Lacking such an analysis, Quine's positivist and Wittgensteinian opponents had no direct way of testing their intuitive idea that certain special sentences were true "by convention." But using Lewis' analysis, we can test it. We can plug the idea of *holding something true* or *treating something as true* into Lewis' analysis and see if the result is plausible. (Lewis' analysis is surely accurate enough to use in such a test, though questions have been raised about its details.[4])

Lewis' formula is as follows (p.78):

A regularity R in the behavior of members of a population P when they are agents in a recurring situation S is a *convention* if and only if it is true that, and it is common knowledge in P that, in almost every instance of S among members of P,

(1) almost everyone conforms to R;

(2) almost everyone expects almost everyone else to conform to R;

(3) almost everyone has approximately the same preferences regarding all possible combinations of actions;

CHAPTER TWELVE

(4)   almost everyone prefers that any one more conform to R, on condition that almost everyone conform to R ["any one more" is a weakening of the stricter "everyone," still in the spirit of "the more the better"];

(5)   almost everyone would prefer that any one more conform to R', on condition that almost everyone conform to R', where R' is some possible regularity in the behavior of members of P in S, such that almost no one in almost any instance of S among members of P could conform both to R' and to R.

Since Lewis has already co-opted the variable "S," let us use "T" to stand for our putatively analytic sentence. For the case of "treating T as true," the "recurring situation S" is either vacuous or a (rare) situation in which one is called upon to assent to T. Now, according to Lewis' analysis, it is a *convention* among members of a population P to treat T as true iff it is true that, and it is common knowledge in P that, in almost every instance of S among members of P: (1) almost everyone treats T as true; (2) almost everyone expects almost everyone else to treat T as true; (3) almost everyone has approximately the same preferences regarding all possible combinations of actions; (4) almost everyone prefers that any one more treat T as true, on condition that almost everyone treat T as true; (5) almost everyone would prefer that any one more treat T otherwise, on condition that almost everyone treat T otherwise, where "otherwise" indicates the possible but unchosen alternative to the actual behavioral regularity that is required for the regularity's being genuinely conventional.

The foregoing statement does not make a great deal of sense. (1) is fine, and (2), (3) and (4) are all right, though somewhat emptily so; but (5) rings true only if understood as supposing that the sentence T might mean something other than what it does mean. We already know that the truth of T is conventional in *that* way. What makes no clear sense is that, given T's meaning what it means, we "would prefer that any one more treat T otherwise [than true], on condition that almost everyone treat T otherwise." Since T, meaning what it means, logically (not: analytically) cannot be otherwise than true and everyone knows that, we do not know what we should prefer if almost everyone were to start treating *T* otherwise than as true. Treating *T* as true is not like driving on the right-hand rather than the left-hand side of the road. Thus a friend of truth-by-convention has a heretofore unanticipated task: to show how Lewis' analysis of convention or some suitable successor applies to the truth of a sentence, in order to make it plausible that that sentence really *is* true *by convention*.

That concludes my Quinean defense of the claim that there is little hope for an account of analytic or "purely conceptual" truth based on stipulative definition. And therefore (if I am right in thinking that the Stipulative Definition account offers our *best* hope) there seems, just at this point, not much hope for an account of such truth at all. Moreover we have reached that conclusion without joining Quine in his radical nihilism about linguistic meaning. The onus is now on the defender of analyticity actually to do some defending.

But in fact, a further case is at hand.

3. THE ANN ARBOR DEFENSE

Paul Boghossian and William Taschek began building the case anew, in the discussion period following presentation of the previous sections of this chapter at the University of Michigan in 1990.[5] Boghossian (1992) gives its present formulation: Suppose, as I insist, that meaning is real and a matter of truth-condition (give and take a bit). A paradigm case of truth-conditional specification of meaning is the stipulative definition of the truth-functional connectives. Let us stipulate in particular that $A \equiv B$ is true iff **A** and **B** share a truth-value, and that this exhausts the meaning of "≡." Then any univocal instance of "$P \equiv P$" is true in virtue of meaning.

If it be protested that no pure instance of "$P \equiv P$" is known to occur in English or any other natural language,[6] Boghossian points out (p. 21) that anyone might augment English with a biconditional connective (say, "biff") whose meaning is specified as follows: Sentences of the form "P biff Q" are true just in case "P" and "Q" express the same meaning. Then every instance of "P biff P" in

Augmented English will be true solely in virtue of its meaning. He concludes that if (as I urge) we are meaning realists at all, then either there are analytic sentences or, because we could so easily manufacture some, there might as well be.

To my disgust, I find Boghossian's argument fairly convincing so far as it goes. Thus I am prepared to give up the thesis that there could not in any language be sentences that are analytic in anything like the sense Quine attacks. Of course, that thesis is a very strong one, certainly stronger than Quine's own (given his startling concession on pp. 25-26 of "Two Dogmas").

(Why have I flirted with the Very Strong Thesis at all, then, in the first place? Because any very robust metaphysical realist ought to blanch at the idea that any sentence could be true, i.e., correspond to reality, by convention, i.e., in virtue of human agreement and nothing more. But at least the metaphysical realist can still have the logical-atomist claim that *atomic* sentences are made true only by substantive correspondence with mind-independent reality; and, as we shall see, the admission of sentences that are "analytic" in the present sense is not particularly antiQuinean--nor does it after all mark such sentences in any clear way as true "by convention.")

Actually, Boghossian's point generalizes,[7] and damages more than just the Very Strong Thesis. Supposing again that meaning is real and codified (give and take a bit) by systematic specifications of truth-condition, it is reasonable to expect that any natural language will contain a significant class of sentences that, in one clear sense, are "true in virtue of meaning." This is because the semantics of any natural language L contains a class, however ill-defined,[8] of operators that act as logical connectives and whose truth-conditional role is specified as part of L's Tarskian truth-definition. (For the truth-functional connectives, at least, their truth-conditional role is all that the truth-definition has to say about them.) Those operators may directly formalize morphemes of L, or they may have no surface reflections but figure only in the semantic underpinning.

In either case they will allow a truth-definition to mark a special class of sentences as *logical truths*; a "logical truth" will be just a sentence that remains true under any uniform reinterpretation of its nonlogical terms. And this embarrasses my Quinean case, by affording a clear sense in which some sentences of L are indeed "true in virtue of meaning": (i) Given any logical truth T of L, T's truth will be deducible from L's truth-definition alone. But (ii) a theory of meaning for L contains a truth-definition for L.[9] It seems reasonable to suppose that (iii) a correct theory of meaning for L, strictly so called, codifies all *and only* the meanings of L's sentences. Therefore (iv) the truth of T follows from L's meaning-facts alone. And since (v) the derivation of T's truth will require only those clauses of L's truth-definition that focus on T's component morphemes in particular, (vi) so there: T's truth is guaranteed by T's own meaning, not just holistically by the semantics of L in its entirety. Call this the "Truth-Definition Argument."

Assuming L's semantics also features some meaning postulates (however arbitrarily those may be distinguished from clauses in L's truth-definition), an even larger class of L's sentences will come out true in virtue of meaning--it will include not just L's logical truths but further consequences of L's truth-definition taken together with L's meaning postulates, such as (in English) the classic "No bachelor is married." This begins to look very bad for the meaning-realist foe of analyticity.

The Truth-Definition Argument's move from (v) to (vi) would require some assumptions, at least one of which would be highly controversial. But I am more concerned to belittle the weaker (iv), and then to dispute the spirit of (iii).

I concede that any truth-conditional semantics for a natural language will yield a subclass of that language's sentences whose truth is guaranteed in the way we have seen by the semantics alone. In that sense, some sentences are "true in virtue of meaning." But finer distinctions are needed here.

In the first place, what has strictly been shown to be entailed by L's truth-definition is only that T is true and so is every other sentence having the same form--nothing about *why* T is true, about T's modal status, or about T's epistemic credentials. Still, the Truth-Definition Argument itself shows (if it is sound) that a complete theory of T's meaning alone entails T's truth, and that ought to make T true in virtue of meaning, and truth-in-virtue-of-meaning is what I myself have insisted we should mean by "analytic." Even though the truth-definition *entails* nothing about why T is true, *its entailing T constitutes* why T is true.

That last point is disputable. As Devitt (1992) observes, that a set of sentences entails a further sentence S does not itself entail that the facts expressed by the set of sentences are what make S true. Only an unreconstructed fan of the Deductive-Nomological theory of explanation would automatically take deducibility as explanation. For all that has been shown, though a truth-definition specifies and codifies meanings and even specifies some truths, it does not explain those truths.

I am not completely convinced by that post-Positivist line of defense, though it is correct so far as it goes; whatever the failings of the Deductive-Nomological theory of scientific explanation, the Truth-Definition Argument still seems to have shown that there are sentences whose truth is *guaranteed by* their truth-conditional meanings. But I do not need the post-Positivist line. I can concede the Deductive-Nomological sense in which some sentences are true in virtue of meaning, those sentences whose truth follows from their containing languages' truth-definitions (call such sentences "TD-analytic"). For the truth-in-virtue-of-meaning thus demonstrated is philosophically innocuous in each of two opposing ways. First, it does not entail alethic necessity of any grade. For all that has been shown, a sentence might be TD-analytic and still metaphysically contingent. A cheap way of seeing this is to note that a theory of meaning for L (if it included a "semantic pragmatics" in Cresswell's [1973] sense as well as a pure truth-definition) might entail that "I exist" is always true. A more expensive way is to recall the possibility that the Law of Excluded Middle is contingent; even if our own world is not one in which the Law fails, there may be worlds featuring suitably different quantum mechanics.

Second, TD-analyticity does not abet the traditional Positivist, antimetaphysical project of *debunking* necessity by trivializing it. The Positivists wanted to show that analytic sentences are contentless, fail to correspond to facts, say nothing "about the world," "merely reflect

STIPULATIVE DEFINITION 277

our habits of usage," etc. (This was the key to their elegant solution of what we now know as Benacerraf's [1973] problem about arithmetical reference. They held arithmetical sentences to be analytic, and therefore that questions of arithmetical reference, arithmetical facts and the like could not arise.) But, so far as has been shown, the logical truths of L (such as "Boghossian and Devitt are philosophers just in case Boghossian and Devitt are philosophers") are still about real things and people, and say perfectly intelligible if boring things about them.

The latter point can be strengthened. When theorists have called a sentence "analytic," meaning "true in virtue of meaning," they have traditionally also meant that "the world" makes no contribution to the sentence's truth, that the sentence holds no matter what the nonlinguistic world is like. But the Truth-Definition Argument does not support analyticity in that sense. Although the truth of sentence T is entailed by our theory of meaning for L, that theory of meaning itself presupposes worldly tautologous facts; and therefore its entailing T all by itself fails to show that the world plays no role in T's truth. If (the real people) Boghossian and Devitt *were not*: philosophers just in case Boghossian and Devitt are philosophers, the tautologous sentence expressing that fact would not be true after all.[10]

To see the dependence more clearly, consider the proof of T from L's truth-definition. Let us take as given the theorem

(BT) "Boghossian and Devitt are philosophers" is true iff Boghossian and Devitt are philosophers.

and the recursive clause

(Biff) ⌜A biff B⌝ is true iff either A and B are both true or A and B are both false.

To derive the out-and-out truth of "Boghossian and Devitt are philosophers biff Boghossian and Devitt are philosophers," we need to detach somehow, both from the supposition that Boghossian and Devitt are (in fact) philosophers and from the supposition that they are not

CHAPTER TWELVE

philosophers. The proof would have to go by vel-Elimination or Dilemma, based on the tautology that Boghossian and Devitt either are or are not philosophers. That tautology would have to be assumed as a premise licensed by our old friend the Law of Excluded Middle. Thus, although we can indeed deduce our logical truth T from L's theory of meaning "alone," the deduction relies on at least one tautology already held to be true. Any use or application of a theory of meaning will in that sense presuppose some laws of logic, and so the world will after all contribute to the derivations of logical truths.

(Someone may point out that in any decent natural-deduction system, Excluded Middle can itself be proved from the null set of premises; so we need not invoke it as an article of faith, in our subsequent derivations from truth-definitions. But this only puts the problem off. The rules of inference employed in the system are validated only by soundness proofs done by truth table, and the reasoning that comprises the soundness proofs itself presupposes Excluded Middle. Perceptive students sometimes complain about this.[11])

Thus, the Truth-Definition Argument's premise (iii) is true only if we do the Argument the courtesy of allowing that the class of "meaning-facts" is closed under entailment and so including the worldly tautologous facts. But the Quinean maintains, to the contrary, that the tautologous facts predate linguistic meaning; so the Quinean has no reason to grant (iii) strictly construed, and will deny (iii)'s "and only" clause. (iv) remains true, but only because of the worldly tautologous facts presupposed by the notion of "following from" that occurs in (iv).

Finally, note that although the truth of a tautologous sentence is explained by the sentence's meaning in the sense I have conceded, the fact recorded by the sentence is a different explanandum and is *not* explained by any truth-definition. No truth-definition explains why the live people Boghossian and Devitt are philosophers just in case they are philosophers; *au contraire*, the truth-definitional explanations of the truth-values of tautologous sentences already presuppose tautologous facts of that sort.

I have reluctantly conceded a sense in which logical truths are "true in virtue of meaning." But that is not a sense in which they are true contentlessly and with no contribution from the nonlinguistic world. Nor is it any help either to the metaphysician in search of immutable necessary truths or to the anti-metaphysician bent on debunking such things.

4. AND SO WHAT?

Moreover, our Ann Arbor concession leaves the foe of analyticity with each of three further fallback positions, all of which I believe are true, interesting, and supported by the barrage of Quinean arguments I have given in the last chapter and this one:

(i) To grant semantic privilege on the basis of (assumed) meaning realism is not to grant any *epistemological* privilege to any sentence of any natural language. For one thing, we may never know whether any existing sentence does reflect "≡." For another, even if certain laws of logic are in fact true and necessarily so, we may not be completely justified in believing them. All the epistemological arguments against unrevisability hold. What Boghossian's point shows is rather that TD-analyticity does not after all require epistemic unrevisability.

Qualification: The point does still require that in some sense, someone who rejects a *de facto* TD-analytic sentence has failed to grasp that sentence's meaning. But we are or should be hard put to say how the difference between uncertainty about logical form and uncertainty about fact should be distributed over epistemology, given that there are Duhemian tradeoffs between the two. I can deny a TD-analytic *sentence*, because I may have reasonable (though in fact misleading) grounds for doubting that the sentence expresses the logically-true proposition it does express. I could even doubt the logically-true proposition, for either of two reasons: First, because, having imperfect access to what proposition it is I am entertaining, I might entertain it without realizing that it is a logical truth. (In this regard, logical truths are no better off than the Kripkean metaphysical necessities discovered by chemistry, or for that matter the necessary but a posteriori identities of persons such as Cicero and Tully.) The second reason is that even if I do know very well what proposition it is I am entertaining, say the Law of Excluded Middle, I may nonetheless have rational epistemic grounds for doubting it; sometimes genuine metaphysical necessities do not carry epistemic credentials to match.[12]

Thus, the new meaning-realist TD-analytic/synthetic distinction has none of the epistemological importance that has been pinned on it by (e.g., Positivist) philosophers.

(ii) Nor, obviously, has TD-analyticity much philosophical importance. In particular, one cannot defend a philosophical premise or other claim by announcing that the claim is analytic and brooking no further dispute. (The same goes for attacking a position one dislikes by calling it "unintelligible," i.e., analytically false, or by asserting that it "makes no sense" or the like.) This is akin to Putnam's (1975d) point that any analytic sentences there may be would be trivial and of no philosophical interest. Even if stipulative definition does yield some analytic truths, they certainly are not the ones that are useful to philosophers, being absurdly trivial. No sentence that has ever been put forward as a necessary truth by a philosopher seriously philosophizing is the product of a stipulative definition. Even when philosophers construct elaborate systems of stipulative definitions, our interest is not in the definitions themselves, which can have no nontrivial consequences, but rather in the way in which the system connects up to the real world, and the latter cannot be stipulative.

(iii) Not even "It is raining ≡ it is raining" is true by convention, if this is taken strictly as meaning conventionally held true (cf. my appeal to Lewis' analysis of convention.) At best its truth is a consequence of the meaning of "≡" conjoined with some linguistic assumptions. Perhaps the assumptions are all individually conventional in some sense; but the notion of "consequence" that figures here again presupposes tautologous facts of the sort that are cited in soundness proofs. Despite my concession of TD-analyticity, we robust metaphysical realists at least are spared the ignominy of truth by convention.

STIPULATIVE DEFINITION

NOTES

[1] I think he would also want to add exactly what he later says about "explication" generally in sections 53 and 54 of Quine (1960): that theoretical "definition" is a replacement or substitution, not an analysis or uncovering of a pre-existing conceptual meaning.

[2] See also Harman, pp. 140-141; Quine's remarks (1963, p. 113) on the rapidity with which stipulative definitions "fade away"; and again Putnam's (1975x) many examples.

[3] An earlier stab was made by Schwayder (1965).

[4] E.g., Jamieson (1975); Burge (1975); Gilbert (1981).

[5] Actually, I had formulated an argument similar to theirs in a 1975 course handout, but then self-servingly forgotten it for fifteen years.

It should be noted that neither party remains in Ann Arbor; Boghossian is now at New York University, Taschek at Ohio State.

[6] One might say this because one believes natural-language conditionals are not truth-functional. I do not myself believe any natural language contains either the horseshoe, any simple strict conditional, or any biconditional made of any of those. (For the arguments, see Chapter III of Lycan [forthcoming].) The semantics of ostensibly biconditional sentences of English is both tricky and infested with the pragmatics of restriction classes and parameter shift. Thus it is not clear that every biconditional, or even every one with ostensibly identical LHS and RHS, is true, much less true by virtue of meaning. (I know that sounds awful. I also deny the validity of Modus Ponens; see Lycan [1992; forthcoming, Ch. III].)

If one tries to make an analytic formula out of a nonconditional connective, the obvious candidate is "or." But (surprise!) I do not think natural languages contain the vel, or any very well-behaved disjunction operator, either.

[7] Boghossian has notified me in correspondence that he himself does not accept this generalization. Nor is it his intention to defend "truth by convention" or to rehabilitate the Positivist theory of necessity and logical truth. So the anti-Positivist points I shall make below are not aimed at Boghossian's own view.

[8] The argument of Chapter 10 does not intrude here, for the point I am about to make does not depend on there being any single best precisification of the notion of a logical constant.

[9] I continue to make this Davidsonian assumption, defended in Lycan (1984). *L*'s theory of meaning would also contain whatever is needed to block Chapter 9's counterexamples to the Modified Intension Thesis, plus whatever anyone thinks should be counted as part of the "meaning" of a sentence over and above its truth-condition.

[10] Yes, I know that sentence is a counterlogical. Unlike David Lewis, I have no problem with counterlogicals.

[11] If it be complained further--say by an inferential semanticist in the tradition of Wittgenstein, Carnap and Sellars--that rules of inference stand on their own as intuitively sound and need no extraneous license from model theory, the Quinean responds with the rhetorical question of what distinguishes the "intuitive" validity of a rule from a conditional belief about the world.

[12] A similar point is made by Devitt (1992).

CHAPTER 13

ANALOGY AND LEXICAL SEMANTICS

Linguistic semantics of the past twenty-five years has been governed by a certain picture. Some people might say that linguistic semantics has been *held captive* by that picture, in Wittgenstein's pejorative phrase; some might even contend in addition that the picture is viciously false. The basic elements of the picture--hereafter, "Lexical Atomism"--are as follows.

(A) [A strong form of the compositionality assumption] The meaning of a sentence is entirely determined by the atomic meanings of that sentence's individual component words (really morphemes) together with the syntactic rules and correlatively the semantic rules according to which the relevant morphemes are combined and arranged into well-formed sentences.

(B) The atomic morpheme-meanings can be given by individual entries in a definitive dictionary of some sort--preferably clauses in a truth-definition either extensional or intensional. These meanings are fixed, by tacit convention, prior to any syntactical combination of the morphemes into longer constructions.

(C) There are only finitely and manageably many morpheme-meanings underlying any single natural language, or for that matter underlying the totality of all extant natural languages, primarily because any single morpheme-meaning in a natural language must be learned individually by any human speaker of that language in a determinate chunk of real time.

(D) Lexical ambiguity in a natural language is neatly limited. Of course a word may have more than one sense, but the several senses it has may be crisply captured in a short, discrete list and treated by linguistic semantics as brutely homonymous or equivocal (otherwise its separate uses could not have been learned in so short a time by speakers).

(E) In addition to its ordinary *literal* meaning or meanings, a word might have some figurative--particularly metaphorical--uses as well. But although both philosophers and linguists rightly count it a great mystery how nonliteral meaning is derived from real meaning, figurative meaning is not only derivative but nearly negligible from the viewpoint of current active linguistic theory. Semantics *proper* studies literal meaning, though we would all admit under pressure that philosophy of language taken more broadly comprehends the question of how nonliteral meanings are generated from literal ones (and how they are related to literal truth-conditions). Metaphor especially is a surd in language that will need very special and arcane explanation once the more straightforward theory of literal meaning or "literal propositional content" has been straightened out.

Several further theses are entailed by, presupposed by, or at least historically associated with the foregoing basic elements:

(F) A word is either *univocal* or *equivocal* (ambiguous, homonymous), though perhaps most words happen to be ambiguous.

(G) A sentence (of a natural language) has at least one determinate *logical form*. Actually, due to manageable lexical ambiguity as well as to equally manageable syntactic or structural ambiguity, almost any sentence has a multiple range of individual "readings" corresponding to *distinct* "logical forms" ascribable to that sentence in different contexts, but that range is not often large.

(H) *The* meaning of a sentence consists in the literal truth-condition(s) expressed by way of its logical form(s); truth-conditions are determined compositionally by literal morpheme-meanings and syntactic-semantic rules (cf. (A) and (B)).

(J) Atomic morpheme-meanings are a matter of important if conventional relations between individual words or morphemes and (predominantly) nonlinguistic items in the world. The paradigms here are *reference* or denotation for singular terms and extensions or satisfaction-classes for predicates.

I do not know of anyone who has articulated the foregoing picture in its entirety, though it is (rightly) associated with the works of Davidson, Hintikka, and Montague.[1] It is a surpassingly lovely picture, and an important one too. But, alas, it is also inaccurate.

I believe Lexical Atomism goes wrong by neglecting the analogy mechanisms at work in lexical semantics, and in particular the volatile polysemy induced by those mechanisms. I was brought to see this (not before time) by reading J.F. Ross' formidable and annoying book *Portraying Analogy* (1981). This chapter is a critical exposition of the portions of that work that I take to spoil the Atomist picture.

1. POLYSEMY

Ross proposes to "look for a systematic account of the meaning relationships different tokens of the same[-spelled] word have to one another" (p. 1). Like Weinreich (1971) and Lyons (1977), he champions an infinite-polysemy thesis.[2] He holds (contrary to (C) and (D)) that virtually any word, even a pronoun, may take on any number of novel and distinct lexical meanings without limit, given a suitable variety of subsentential linguistic environments, and moreover in such a way that the novel meanings can be grasped on the spot by normal hearers. This is because novel word-meanings are generated in context from existing ones by intricate but quite tractable mechanisms of "analogy" that are mobilized by all normal speakers. And for the same reason, importantly, very little difference of word-meaning is "mere" or brute equivocation (contra (F)); the polysemous meanings are systematically interrelated.

If it should be replied *á la* (E) that this is uninteresting even if true since "merely" figurative and particularly metaphorical usages are cheap, Ross responds by powerfully attacking both the distinctions presupposed by that reply: the distinction between (what we ordinarily take to be) literal meaning and "figurative" meaning, and the distinction between univocity and "mere equivocation" or homonymy (cf. (F)). Almost all word-meaning is derivative or (as we would say) "nonliteral" to some degree. The "analogy phenomena" in particular, including analogy of proportionality, simple metaphor, denominative analogy and paronymy, are ubiquitous.

The analogy mechanisms all involve what Ross calls "differentiation": a word differentiates when it is forced to take on different meanings in order to adapt to different subsentential environments, normally to other words with which it is concatenated. A given word differentiates in context because the other expression "dominates" it in that context.[3] (Some words are easily dominated, "dominated in a great variety of ways, especially common words like 'make', completed with 'time/trouble/way/appointment/bed/money/merry/haste/cake/dinner/love/war'" (p. 9); those are the words having the most diffuse variety of possible meanings. Other words are comparatively "intransigent" and, in practice at least, have very few meanings, though "all words are dominated some of the time.") The direct cause of differentiation is the interplay of "linguistic inertia" and "linguistic force." The first of those is the principle that two tokens of the same word have the same meaning unless something differentiates them; the second is the principle that

words resist combining unacceptably in the linguistic environment, until forced to.... In other words grammatical strings will not go together unacceptably...if there is any step-wise adaptation of word meanings (comparatively to their other occurrences in the corpus) which would result in an acceptable utterance and is not prevented by the environment. And those step-wise adaptations are the specific kinds of differentiation described in this book. (p. 10)

Thus when two words combine and one of them dominates the other, the dominated word adapts if adaptation is required in order for the resulting phrase to be "acceptable." "Sentences make what sense they can" (p. 11).

By "acceptable" Ross does not mean merely *grammatically well-formed*; the latter is by far the more inclusive notion. But he gives no general characterization of unacceptability, even though it is the most basic idea in his scheme; he is content to note that it is "variously based" (pp. 10, 81), to deny that logical inconsistency suffices for it (p. 82), to allude to "computability" on the part of "competent writer-speakers" (p. 58), and to certify that sentences may be "unacceptable" simply in that they are obviously false (p. 11) or for that matter too obviously true (p. 114). We are to get the idea more firmly from the book's profusion of examples.

Speaking of examples, it is more than time for some here. To begin, I shall just list a few cases of fairly wide differentiation *per se*: (i) "She *dropped* a stitch"; "She *dropped* her hem-line; "She *dropped* her book"; "She *dropped* a friend"; "She *dropped* her courses" (p. 33). Ross maintains that each occurrence of "dropped" in this list means something different. Moreover, "[t]he meanings...are appropriate, *fitted* to the completion words...." (ii) "He picked a date"; "He appointed a date"; "He fixed a date"; "He wanted a date"; "He borrowed a date" (pp. 80-81). Ross notes that these sentences are still ambiguous, and the ambiguity could be reduced only by the addition of wider contexts. (iii) "He *charged* the gun"; "He *charged* the jury"; "He *charged* her with murder"; "He *charged* him with responsibility"; "He *charged* more than the law allowed"; "He *charged* the boy too much"; "He *charged* the battery";... (paraphrased from p. 100). (iv) "Dead man"; "dead duck"; "dead silence"; "dead ringer"; "dead march"; "dead eye"; "dead end"; "dead head"; "dead assets"; "dead heat"; "dead bolt"; "dead language"; "dead wrong"; "dead drunk";

"dead tired," "dead boring"; "dead set (on)"; "the dead of winter." (My own example, offered in the spirit of Ross' view. Ross here and there observes, triumphantly but correctly, that once one is aware of the phenomenon one sees it everywhere.) (v) Prepositions such as "in" and "on" are *notorious* differentiators, having virtually no constant meaning from context to context. (Ross himself does not stress prepositions, but Lakoff and Johnson [1980] make the point trenchantly.)

The reader will have gathered that Ross individuates meanings very finely; one might well deny that *all* the foregoing word uses differ in meaning. Accordingly, in section 3 of Chapter 2, Ross offers a catalogue of no less than twenty ambiguity tests, culled from Aristotle. The tests are eminently disputable, but he does not claim either completeness or decisiveness for them; he aims only at "a kind of reflective equilibrium reached through the feedback adjustments of both tests and intuitions" (p. 40). (Oddly, he makes no appeal to the tests proposed in Zwicky and Sadock [1975]; nor does he consider the method of asking whether a sentence containing a term under investigation can simultaneously have more than one truth-value. Had he done so, he might have armed himself with a more incisive instrument for the defense of his ambiguity intuitions.)

2. ROSS' APPARATUS

In Chapter 3 Ross begins the serious and very turgid business of defining the technical terms that figure in his explanatory apparatus.[4] It is here (Def. *III-11*, p. 82) that "dominance" is explicated, in particular, and readied for use later on in the explanation of various specific kinds of differentiation. Ross' system of definitions is exhaustingly dense, with heavy jargon muliplying exponentially throughout the chapter (and beyond),[5] but I shall try to limn a bit of the strategy though omitting quite a lot.

Ross' initial idea (following Lyons [1963] and ultimately the structuralists) is that

[t]o have meaning, an expression must be *in contrast of meaning*; it must have *differential* meaning--function contrastively to something else.... An expression *paradigmatically* contrasts with all other expressions that could have been inserted in its place [in a sentence whose remainder stays] substantially unchanged...to produce replacement sentences that are semantically *acceptable* within the given environment and preserve the consistency or inconsistency [n.b., *not* necessarily the truth-value] of the original. (p. 54)

The set of all those other "co-applicable" expressions would be called the "paradigmatic set" for the original expression in its context. (E.g., "The youth/young man/boy/urchin/ girl/matron/etc. asked for a job.")

A paradigmatic set begins to approximate to the sort of pattern of "affinities and oppositions" that Ross thinks makes for lexical meaning. However, we cannot just go ahead and identify a word's meaning in context with its paradigmatic set, for "[paradigmatic] sets may be the same when the tokens differ in meaning and...the sets may differ when the tokens are the same in meaning" (p. 63). Ross wants to say that only some of the members of an expression's paradigmatic set are "meaning-relevant" to that expression. Very roughly, a member *is* meaning-relevant to it just in case, though it is co-applicable in the given context, it is in at least one other context *not* co-applicable with it, "where the not being co-applicable (or not being in the same affinity or opposition of meaning) *marks* a meaning difference" (p. 66).

This last quoted qualification is crucial. Consider (p. 67) "When Smith first married, he married his first cousin." "His aunt's son" can be substituted for "his first cousin" when "married" means "performed a marriage," but not when it means "entered into marriage"; yet this indicates no difference in the meaning of "cousin" across the two types of occurrence, but is due precisely to the different meanings of "married," and so the difference of paradigmatic set for "cousin" in the second case does not suffice for difference of meaning.

Once the foregoing qualification is observed, however, Ross contends that the meaning-relevant co-applicables of an expression in a context can be isolated. He says they turn out to be, "[o]n the whole, the *near synonyms*, the *contraries*, the *determinables*, the *determinates*, and the *predicates of correct contextual definitions*...."

And *this* set of "affinities and oppositions" does seem to be a plausible candidate for specifying lexical meaning.

Consequently Ross introduces the notion of a "predicate scheme" (p. 63), narrower than that of a paradigmatic set. An expression's predicate scheme in a context is that proper subset of its paradigmatic set whose members are meaning-relevant in the sense defined.[6] Out of context, of course, an expression belongs to each of many different predicate schemes; context reduces ambiguity by narrowing down the number of schemes.

> Where there is a scheme difference for the same word, the contrast between the two completion expressions...selects the scheme, e.g., 'wanted': 'The beggar wanted money'; 'The bearing wanted oil'. The selection process is *exclusion* through the resistance (intransigence) of the completion expression to concatenating with any scheme for 'wanted', that contains any word that, substituted for 'wanted', would yield an unacceptable utterance. (p. 81)

The second of these two quoted sentences summarizes "dominance"; the completion expression dominates the term whose scheme-indifference is thus reduced. (N.b., dominance is not antisymmetrical; each of two concatenated expressions can dominate the other so long as they dominate in different respects (pp. 51-52).) An example:

> 'Pen' is indifferent to *instrument* schemes (that include 'pencil', 'quill', 'stylus', etc.) and to *enclosure* schemes (that include 'stall', 'paddock', 'run', etc.). The verb 'wrote' is indifferent to *means expressing* schemes (that include 'telephoned', 'telegraphed', 'semaphored with', 'drew with', 'inscribed with', 'signed with', etc.). Yet, concatenated into 'He wrote with a pen', 'wrote' is *not* indifferent to the *enclosure* scheme for 'pen'; rather, it is *resistant* because it belongs to a *means demanding* scheme which, concatenated with that enclosure scheme, would in the supposed environment yield an unacceptable sentence, e.g. 'He wrote with a barnyard enclosure'. 'Wrote' dominates 'pen'. And 'pen', thus dominated, reciprocally dominates 'wrote' to exclude the *means expressing* scheme ('telegraphed', etc.) because that result would be similarly unacceptable. (p. 86)

The explanatory apparatus as described gives off a strong air of circularity, and I am not sure it has genuine explanatory power.[7] But here I am concerned only with taxonomy and with the descriptive attack on Lexical Atomism, so we may continue without worrying over explanatory power.

Ross now sets out (Chapters 4 and 5) to apply his apparatus to the "analogy phenomena," thereby to show how mere equivocation, analogy of proportionality, metaphor, denominative analogy and the like differ from each other and how in particular they are differently caused by dominance. I shall discuss one illustrative case, that of simple metaphor, both for illustrative purposes and because metaphor is probably the most interesting of the analogy phenomena to workers in the rank and file of linguistics and the philosophy of language.

3. METAPHOR

Ross initially characterizes "simple metaphor" as a meaning relationship satisfying the following description:

[S]ame words, taken in pairs, that differ but are related in meaning. Some near synonyms...of the one instance are also near synonyms for the other. Further, the difference in their meanings is marked by a difference in the group of meaning-related *other* words that can be substituted for the two instances.... [They] also satisfy an additional condition of asymmetry. Roughly, certain meaning-related words substitutable for the non-metaphorical occurrence are not substitutable for the metaphorical occurrence, even though other words 'implied' by those very words *are* substitutable and are part of what is meant. (p. 7)

Simple-metaphorical occurrence is an obvious case of dominance and "linguistic force," since the unacceptability in some sense of the literal reading is proverbially what enforces the metaphorical reading. And Ross reminds us, as is his custom, that this process is automatic, a dynamic feature of language itself. "Metaphors don't have to be made, they can just happen as words fit their contexts" (p. 113).

Ross starts with Goodman's (1968) idea that "a metaphor is created when a word belonging to a label scheme or classificatory

CHAPTER THIRTEEN

scheme...that has a certain prior realm of application...is subsequently applied to something not belonging to that realm and, by implication, segments the realm into ranges of referents corresponding more or less well with the contrasts among the original label scheme" (p. 110). But Ross complains that Goodman's idea explains neither "what the contrasts within the label scheme consist in," nor "how those contrasts are preserved in the application to a new realm," nor why the metaphor relation is asymmetrical (pp. 110-111). According to Ross, "[t]he asymmetry is created by the mandating of a *double* presupposition, an 'implying' and an 'implied' presupposition in the literal token, and by the mandating [*via* dominance] of the (relatively) 'implied' presupposition in the metaphorical token and the *exclusion* of the 'implying' contrast" (p. 112). This needs a bit of explanation.

The members of a predicate scheme contrast with each other. To take Ross' own example, the word "swept" contrasts with other cleaning terms such as "washed," "polished," "mopped," "scrubbed" and "dusted." Each action verb in this scheme implies a corresponding physical motion, a corresponding apparent motion, a corresponding effect, and the like. The physical-motion verbs contrast with each other in a way parallel to that in which the original action verbs do; so do the apparent-motion verbs, the effect descriptions, and so on. Thus, according to Ross, the original contrast in the predicate scheme of cleaning terms as a whole implies each of the other contrasts. Now, simple metaphor occurs when a subsentential environment applying to a word such as "swept" excludes (on pain of unacceptability) one presupposition of its intrascheme contrast, such as that of sweeping's being the act of a living person, but "mandates" or enforces one or more of the implied contrasts. Thus "The clouds swept the moon," "His gaze swept the crowd," etc. Ross maintains that this feature accounts for the asymmetrical character of the metaphor relation, over and above the mere fact of differentiation enforced by unacceptability.

Ross denies (p. 114) that metaphors need be literally false. Some are, when the unacceptability that causes differentiation *is* egregious falsity. When the unacceptability is rather "linguistic nonsense" of the category-mistake sort (assuming there is any such thing as opposed to mere egregious falsity), the literal reading if any will be truth-valueless. However, when the unacceptability takes the form of *triviality*, as Ross says it can be ("He's a thing"), the literal usage is true. It all depends.

Some writers use "figurative" as a synonym or near-synonym for "metaphorical." Ross declines to join them, for he thinks that simple metaphor *per se* is just one of several analogy phenomena albeit a distinctive one, and is no "less literal" than is, say, analogy of proportionality; moreover, few of the standard "figures of speech" require *metaphor* at all.[8] He offers (in Chapter 6) his own analysis of "figurative" or "heightened" discourse that presupposes metaphor but goes beyond it. What is distinctive of such discourse, he says, is stepwise *double* differentiation involving both metaphor and analogy of other sorts. Here is an example, accounting for the figurativeness of the verb in "The critics paddled the directors" as it occurs in a poem about a play's closing:

Denominative differentiation of the name of a process (to paddle) to name an instrument (a paddle). The instrument name for a kind of tool (a paddle) is abstractively differentiated into an instrument not a tool (a ping-pong paddle). Then by metaphor an instrument name for a game (a ping-pong paddle) is made into a name for a *weapon*, so that 'initiates are paddled'. Next, a reverse denomination, where the name for the instrument is differentiated into the name for the process of using the weapon, 'to paddle the initiates'. Finally, by analogy the process name is differentiated abstractively to mean 'punish', in the sense of 'inflict deserved condemnation', so 'the critics paddled [the] directors' by inflicting deserved castigation. Through this somewhat fanciful sequence, we have adaptations that involve: denomination, metaphor, denomination, and abstraction. (p. 147)

Ross' "general hypothesis" is that figurative and other complex meaning-relatedness can always be "decomposed into stepwise atomic adaptations...whose structures are ubiquitous and law-like in the corpus of natural language."

I am not sure how the one metaphorical step in the foregoing sequence of adaptations fits Ross' earlier characterization of metaphorical asymmetry. The original contrast would be between ping-pong paddles and other sporting instruments such as bats, racquets, golf clubs and so on. That contrast "implies" a contrast between various different arm motions, a contrast between various different kinds of projectiles, and the like. Does "paddle" differentiate into the name of a weapon by the exclusion of its presupposition of being used for recreation, or on inanimate objects? (Weapons can be used for recreation, or on inanimate objects.) Moreover, what "implied" contrast is mandated by the newly metaphorical "weapon" use of "paddle"? I suppose that between the arm motions, since paddling a person involves a different arm motion from beating or clubbing, but I am not sure there is a fully parallel contrast set corresponding to all the different kinds of sporting equipment that would inhabit the predicate scheme for "paddle." This sort of discussion must be carried on across a broad range of examples of simple metaphor, to see if they all do fit Ross' pattern, in order for his theory of metaphor to face and pass empirical testing. So far, the theory is a daring and untried though not implausible conjecture.

4. TWO CRITICISMS

It is time to note two more (quite tendentious) features of Ross' overall view, and to record a few objections.

First, Ross is no friend of Grice; he is an *anti-intentionalist*. Over and over he stresses, "[a]t the risk of being exasperating" (p. 118), that the adaptations and other dynamic linguistic phenomena he talks about are "automatic, irresistible, non-intentional, law-like and clearly intrinsic to the linguistic expressions in which they occur." "Writing and speech are not encodings of one another or of something in the mind" (p. 22); rather, *au contraire*, speech is *itself a medium of thought*, and fortunately so, for in this way the analogy mechanisms are able to

do part of our thinking for us. Now we can think *in* the distinctions of word meaning without having, continuously, to think *of* the distinctions of word meaning. (p. 100)

ANALOGY AND LEXICAL SEMANTICS 295

Language has a causal structure of its own, independently of what anyone thinks about it or what anyone's intentions are.

I sympathize with this, having defended a related thesis myself (Lycan [1984]). But the closest thing to an argument Ross gives for it is to say (p. 11) that "[t]here simply could not be" infinite polysemy based on linguistic force and so on "if *linguistic meaning* were not something intrinsic to the language and distinct from speaker-meaning...." Why not? Granted no speaker could intend all that infinite polysemy or even more than a minuscule fraction of all the differentiation that occurs in her speech every day, it still might be the case that linguistic meaning is parasitic on speaker-meaning, for speakers' internal systems of representation may exhibit analogy mechanisms and infinite polysemy as well. Might thinkers not differentiate silently, without vocalizing? Further, we know that reference to speaker-meaning cannot be expunged from our theory of language, since virtually no sentence is ever fully unambiguous on an occasion of its utterance, and hearers still have to compute speakers' intentions in the course of disambiguating in communication (I do not know whether Ross means to deny this).

The second tendentious feature of Ross' view disturbs me a good deal more. It is a kind of general linguistic solipsism, hand-in-hand with his anti-intentionalism: He contends that lexical meaning has no more to do with *reference* or any other word-world relation than it has to do with speakers' intentions.

It has too long gone unchallenged that what words 'man' concatenates with acceptably...*depends* upon what it stands for.... Is it not obvious that what 'man' stands for is affected (perhaps, determined) by what words it concatenates with (in a given environment)? It is time to disentangle 'reference' and 'standing for' from some kind of labeling and consider it as another mode of linguistic meaning. (p. 135)

Moreover Ross denies (p. 48) that words out of context already have "context-neutral kernel[s] of meaning" that adjust or adapt when the words are combined. "Outside a context a word has no signification at all (nor any denotation)...." Lexical meaning is *exhausted* by "patterns of affinities and oppositions...to other words" and "can be [adequately] represented as the pattern of such relationships" (p. 107).

Here as before, the argument is only that "[t]here simply could not be" analogy mechanisms of the sort Ross' book is about if lexical meaning were even in part a matter of reference. But here it is even less obvious why we should grant that assertion. (Perhaps Ross is thinking that if meaning were in part the determination of reference, then for a word to differentiate its utterer would have to check on and manipulate its referent in some way, but that is not true either.) And there are several strong reasons not to share this structuralist disdain for word-world connections.

First, note that differentiation presupposes an antecedent meaning to differentiate from. On pain of regress, there must be "absolute," pre-differentiation word meanings. Ross is aware of this but does not care. His book is simply not about absolute meaning. He complains that theories of absolute meaning

do not account for (and usually do not recognize) the analogy phenomena.... I venture that absolute meaning is to a far greater extent dependent upon antecedent relative meaning...than the reverse, and that the idea that primitive humans endowed grunts and twitters with ranges of referents (and subsequently with rules of combination) is pure superstition. (p. 14)

Yet this ignores the regress. Meaning that is "dependent upon antecedent relative meaning" *is* relative meaning, not absolute. What are we to say of absolute meaning? The field is open to those of us who would identify it with a well-behaved extensional or intensional word-world relation.

Yes, Ross might concede, but the field is also still open to his structuralist idea. Granting that there is absolute meaning that is in no way the product of differentiation and the analogy mechanisms, it still may be entirely a matter of "affinities and oppositions," word-word rather than word-world relations; the existence of absolute meanings does not refute that hypothesis.

Perhaps not. But it is a good deal harder to see how primitive people would have established complex patterns of affinities and oppositions than it is to see how they would have endowed grunts and twitters with ranges of referents. And there are at least two further objections to Ross' linguistic solipsism.[9] One is that meaning plus fact determines truth: If you know the meaning of a sentence and you are omniscient regarding fact, then you know the sentence's truth-value. Referential semantic theories explain that fact straightforwardly, by assigning truth-conditions to sentences, compositionally out of relations of terms to their extensions according to various compounding rules. How might Ross explain it without appeal to extensions? Indeed, it is hard to see what his view has to do with truth at all, or how he might even begin to see truth as determined by meaning plus fact.

My remaining objection invokes a "Twin Earth" hypotheses in the style of Putnam (1975e). Suppose that somewhere in outer space there is a planet just like ours, molecule for molecule, running along exactly in parallel with our own Earth (its water is even H_2O, unlike Putnam's own Twin water). Its inhabitants speak a language that sounds just like Earth English. Let us call the other planet "New Earth," in honor of the only difference between it and Earth proper: it came into existence just a few days ago, complete with memories, records, geologic strata and so on.

As I said, the language spoken by the denizens of New Earth-- New Lycan, New Ross, New Putnam *et al.*--is just like English in all its "intrinsic" features. In particular, the patterns of affinities and oppositions that obtain between expressions of New English are exactly like those that obtain between the corresponding expressions of English. It follows from Ross' view that the expressions of New English have just the same meanings as their English counterparts. Yet whenever the inhabitants of New Earth speak of their past (simultaneously and in parallel with our own utterances about Earth history), their assertions are false or truth-valueless; New Babylon, New Frege, New Chernobyl and so on never existed. Now, how could two sentences having just the same timeless meanings differ in truth-value? Moreover, it is fairly clear why corresponding English and New English sentences do differ in truth-value: it is because their terms differ in their extensions. We Earthlings succeed in referring to Babylon, Frege and Chernobyl, while (through no fault of theirs) the New Earthlings do not often succeed in referring at all.

I conclude that Ross' view is seriously deficient, and that we have powerful motives for reintroducing referential relations into lexical semantics.

Nonetheless we can separate the defective component of Ross' theory fairly sharply from what I am persuaded is its genuine contribution. For his linguistic solipsism, and even his (so I claimed) potentially circular account of dominance, are independent of his ideas about infinite polysemy and its origin in differentiation driven by dominance. Those ideas may be perfectly true, even if some quite different explication of dominance is needed and even if we grant that reference is required as an element of lexical meaning. Indeed, Ross has fairly well convinced me, and against my will, that they are true, and that Lexical Atomism needs re-examination. So let us assume their truth for the sake of argument, and see what conclusions should be drawn.

5. REPERCUSSIONS AND ALLEGED REPERCUSSIONS

I begin with Ross' two remaining chapters. In Chapter 7 he is concerned to defend the meaningfulness of religious language against the charge that religious assertions stretch the uses of anthropomorphic terms so far out of their normal boundaries that they cannot have their normal meanings and that even their analogical uses are too attenuated to be literally meaningful. (Needless to say, if a religious assertion is not literally meaningful then it is not literally true either.) Ross argues, predictably, that analogicalness does not entail mere equivocation, and that if chains of stepwise adaptation connect religious uses of terms to less exotic uses, there is no *a priori* reason to insist that religious utterances are "nonliteral" or otherwise to be patronized--at least no more so than a great many other sorts of utterances whose truth we are entirely comfortable with. I find this point entirely convincing.

Similar points can usefully be made in areas other than the philosophy of religion as well, for the inference from "cannot have its normal meaning" to "is meaningless or at best cognitively pointless" is not uncommon in philosophy of mind, in philosophy of science, in ethical theory and elsewhere. To mention one example: Scientific anti-realists sometimes note that key theoretical terms change their meanings in the course of theory replacement, and infer from this that the successor theory is not really a competitor of the replaced theory but is rather incommensurable with it.[10] But this assumes that all amphiboly is mere equivocation, and discounts close analogical relations between successor and predecessor terms. To understand the semantics of theory replacement, one must approach case studies such as that of Newtonian mass and relativistic mass with a theory of analogical extension in hand.

To stick with philosophy of science for a moment, I think a theory of analogical extension would also help with the puzzle of parameter addition in general. In theorizing and in ordinary life, we often add argument places to predicates whose -adicities we thought we knew; sometimes we do this for our own purposes, while sometimes we claim to be discovering a parameter that was hidden in the predicate all along. (Some vexed cases: Consider "5:30"; "simultaneous"; "winter"; "polite"; "wrong"; "true.") This is puzzling, for on the one hand it seems, as writes Wallace (1975, p. 54), that "no more fundamental change can be made in a theory than changing the number of argument places of one of its key concepts"; and one would think the same would go for relational predicates in everyday discourse. On the other hand, we often do not even know how many argument places our underlying morphemes carry, and may never find out, yet our use of the language is not at all flawed by that; how, then, can the exact -adicity be of any consequence? I have suggested elsewhere (Lycan [1989]) that parameter addition should be seen as analogical deformation, though I am far from having a theory of analogical extension myself.

Thus I believe Ross has done well to point us in the analogical direction for understanding lexical semantics as it figures in some philosophical problems.

However, in Chapter 8 Ross attacks truth-conditional theories of meaning, and that I do not find convincing at all. Naturally Ross accuses truth-conditional theorists from Frege to the present of ignoring the analogy mechanisms, and he is right, but this in itself does nothing to show that what the truth-conditional theorists did say is false. We need to see arguments leading from premises about the analogy phenomena to conclusions that impugn the truth-conditional program. Ross does sketch a few such arguments; I shall comment *seriatim*.

(i) It is unclear what truth-conditional theorists mean by "knowing a sentence's truth-condition," and at least one ontological interpretation of that slogan (due to Wiggins [1971]) runs into an unhappy dilemma (pp. 181-182). *Response*: Yes, one has to be more careful in stating one's epistemic thesis than Wiggins was in the article Ross has chosen to attack. I for one have tried to get this right, in Lycan (1984).

(ii) "*Meaning is inherent in English sentences, not attached, as to a code*" (p. 183). Ross seems to think this cuts against truth-conditional semantics. *Response*: That depends on just what it is that Ross is denying. Sentences have the truth-conditions they do regardless of any one person's or group's beliefs or intentions, just as Ross says words have their affinities and oppositions. Nonetheless meaning is *conventional*, in that but for the existence of social practices generally, however diffusely characterized, a string of words might have meant something other than what it does mean, or have meant nothing at all; if in the statement quoted above Ross intended to deny that truism, then he is simply wrong.

(iii) Sentential meaning cannot be identified with truth-condition because sentences can be used to perform illocutionary acts other than that of stating or asserting, and because deictic sentences make different statements on different utterance occasions (p. 184). *Response*: Both premises are true, and both have caused well-known problems for the truth-conditional approach. However, the problems have proven to be tractable, and are by now rather well in hand; see, e.g., Lewis (1972), Lewis (1980), and Chapters 3 and 6 of Lycan (1984).

(iv) Knowledge of truth-conditions is insufficient for knowledge of meaning, since I can know the truth-condition of a sentence that I do not understand. *Response*: Of course I agree, having defended that claim myself in Chapter 9. It does not, of course, count against the thesis that knowledge of truth-condition is *necessary* for knowledge of meaning. I am still unsure what further condition might be needed.

(v) Truth-conditional theorists presuppose a level of "logical form" that is abstracted away from a sentence's surface-grammar and from apparent word meanings. But once the analogy phenomena are properly respected, "logical form" goes relative--or at least, the required assumption of univocity of the "logical" expressions that define "forms" fails more often than it holds (pp. 188-189).[11]
Response: The truth-conditional project per se is unscathed by this, since in speaking of valid inference-patterns one always assumes *by hypothesis* that the words involved are univocal, however seldom words may occur univocally in real life. On the other hand, Ross is right to warn us that the use of truth-conditional methods in the actual study of texts would be seriously hampered by the ubiquity of the analogy phenomena.

I do not see that Ross' arguments have done much damage to the truth-conditional program. But I think they have done considerably more damage to the Lexical Atomist picture. I shall close this chapter by reviewing the tenets of Lexical Atomism and seeing how much of the picture survives contact with infinite polysemy and the analogy phenomena.

Thesis (A) still comes out true, so far as I can see; even if every word in a sentence has differentiated all over creation, the meaning of the whole sentence is still determined by the (however analogical) meanings of the words together with their mode of compounding. But this affirmation of (A) must be qualified by a denial of (B). Only the most common atomic morpheme-meanings can be listed in a dictionary; other perfectly possible meanings will always be left out. And most ordinary meanings cannot be fixed prior to syntactical combination of the morphemes into longer constructions. (Though if there are absolute meanings, these can be so fixed.)

(C) becomes equivocal. If polysemy is infinite, then *a fortiori* there are not only finitely many morpheme-meanings underlying a natural language. On the other hand, the polysemous meanings are derivative, being projected in regular and systematic ways by way of the analogy mechanisms, and so learnability is saved; the mechanisms must work upon a finite base, for all the familiar Davidsonian reasons. (This constitutes a second argument for the existence of absolute meanings.)

(D) and (E) are just false, however nice a world this would be if they were true.

(F) is true in letter, but false as intended. Univocality is extremely rare if it exists at all, and the differences between the different types of equivocity are more interesting and important than is the difference between any of them and univocity. (G) is subject to the qualification noted in my comment on argument (v) above, and the "readings" associated with an ambiguous sentence would either have to abstract away from word meanings entirely or would be far more various than (G) allows. (H), of course, omits mention of the analogy mechanisms, and so is false as it stands, since the analogy mechanisms almost invariably figure in what we would unthinkingly have called the "literal" meanings of sentences.

(J), finally, is immune to the analogy mechanisms themselves, so long as we are still withholding our assent to Ross' particular explication of dominance and his solipsistic view of word meaning. It is only if we were to agree on the latter that (J) would have to go.

Thus there is not a lot left of Lexical Atomism, even though truth-conditional semantics itself is still aloft. I take this to show that the two are largely independent, even though most truth-conditional theorists have tacitly accepted Atomism because they did not particularly care about lexical semantics. But the apparent falsity of (B) and (E), at least, should make semanticists take notice.

NOTES

[1] See Davidson (1967b), and the various essays collected in Davidson (1984); Hintikka (1969a); Montague (1974).

[2] Cf., subsequently, Cohen (1985), and Davidson (1986).

[3] It should be clear that Ross uses the word "context" to mean, not context of utterance in the more usual sense, but subsentential environment--the other word or phrase with which the given word is concatenated.

[4] Ross is fond of insisting that his explanations are *causal* explanations in as robust a sense of causality as one likes (e.g., pp. 5, 49, 53; this provocative claim is the final flourish on what I shall be calling his anti-intentionalism--see below). Some philosophers will find this silly; I rather like it.
[5] Some of the flavor will be conveyed if I quote just the definition of dominance itself:

F *dominates* T^I in C^I (*in E*): (a) if a given term, t, is indifferent, apart from context, to a number of schemes, and (b) if a given frame expression, F (for the sentence C^1 in E that contains t) is semantically (syntagmatically) resistant to (Def. III-12) *some* members of some of the schemes for t, then F dominates the scheme-indifferent t by restricting t's range of schemes for the given context to schemes all of whose members are co-applicable (Def. III-9) with F (if there are any).

Portraying Analogy will never be accused of frivolousness.
[6] The notion is cognate with what some theorists have called a "lexical field."
[7] Difference of meaning is explained in terms of dominance; dominance is defined in terms of predicate schemes; the notion of a predicate scheme is obtained by adding the meaning-relevance condition to paradigmatic set membership. But meaning-relevance is defined squarely by reference to difference of meaning (*Def III-9*, p. 68): to quote Ross' own gloss on the official definition, "[t]he meaning-relevant words are the ones whose *not* being co-applicable somewhere else *marks* a difference of predicate meaning" (p. 67). So I cannot for now discern the source of the whole apparatus' explanatory power. There is no obviously recursive element in any of the definitions, so Ross is not specifying one kind of meaning difference and defining or explaining other kinds derivatively by reference to the first kind.

On the other hand, the offical definition includes the phrase "on serviceable tests for equivocation," alluding, we may gather, to the twenty Aristotelian ambiguity tests I mentioned earlier. Perhaps "basic meaning difference" could be operationally defined in terms of the tests, and then differentiation of various special kinds could after all be defined and explained in terms of basic meaning difference. Yet (a) the tests are supposed to be more or less adequate to our intuitive judgments of ambiguity, not operational definitions, and (b) one feels that the success of the tests should be explained by genuine meaning difference, not the other way around. Or perhaps Ross' claim is only that if we take for granted an intuitive sense of meaning difference, constrained in "reflective equilibrium" with the ambiguity tests, we can by his means expand it into a finely-tuned instrument for at least *predicting* all the analogy phenomena. Yet if predicate schemes are just deliberately mocked up to match intuitive differences in meaning, then explanation of differentiation in terms of dominance still seems a very shallow explanation.

[8] Of all the classical figures of speech, Ross argues later on (p. 152), only personification *strictly* requires metaphor.

[9] These echo two objections I made in Lycan (1984) against the application of Putnamian "methodological solipsism" to linguistic semantics.

[10] Kuhn (1970) has been interpreted as arguing in this way.

[11] Incidentally, Ross has some very illuminating things to say about the use of "conceptual analysis" in philosophy and about the method of chisholming.

BIBLIOGRAPHY

Ackerman, D.F., 1979a, "Proper Names, Essences and Intuitive Beliefs." *Theory and Decision* Vol 11, pp. 5-26.

———, 1979b, "Proper Names, Propositional Attitudes and Nondescriptive Connotations." *Philosophical Studies* Vol 35, pp. 55-69.

Adams, R.M., 1974, "Theories of Actuality." *Noûs* Vol 8, pp. 211-231; reprinted in Loux (1979).

———, 1979, "Primitive Thisness and Primitive Identity." *Journal of Philosophy* Vol 76, pp. 5-26.

———, 1981, "Actualism and Thisness." *Synthese* Vol 49, pp. 3-41.

Almog, J., 1984, "Semantic Anthropology." In French, Uehling and Wettstein (1984), pp. 479-89.

———, 1985, "Form and Content." *Noûs* Vol 19, pp. 603-16.

Amis, K., 1991, *Memoirs*. New York, Summit Books.

Anderson, A.R., and N. Belnap, 1975, *Entailment*, Vol. 1. Princeton, NJ, Princeton University Press.

Armstrong, D.M, 1968, *A Materialist Theory of the Mind*. London, Routledge and Kegan Paul.

———, 1978, *Universals and Scientific Realism*, 2 Vols. Cambridge, Cambridge University Press.

———, 1981, *The Nature of Mind and Other Essays*. Ithaca, NY, Cornell University Press.

———, 1983, *What is a Law of Nature?* Cambridge, Cambridge University Press.

———, 1989a, *A Combinatorial Theory of Possibility*. Cambridge, Cambridge University Press.

———, 1989b, *Universals*. Boulder, CO, Westview Press.

———, 1991, "Classes are States of Affairs." *Mind* Vol 100, pp. 189-200.

Arruda, A., 1979, "A Survey of Paraconsistent Logic." *Proceedings of the Fourth Latin American Symposium on Mathematical Logic*. Amsterdam, North-Holland.

Aune, B., 1967, "Must." *The Encyclopedia of Philosophy, Vol. 5: Logic to Orobio* (ed. P. Edwards), New York, Macmillan, pp. 414-16.

Austin, D., 1983, "Plantinga's Theory of Proper Names." *Notre Dame Journal of Formal Logic* Vol 24, pp. 115-32.

Baker, L.R., 1982, "Underprivileged Access." *Noûs* Vol **16**, pp. 227-41.
Barwise, J., and J. Etchemendy, *The Liar*. Oxford, Oxford University Press.
Benacerraf, P., 1973, "Mathematical Truth." *Journal of Philosophy* Vol **70**, pp. 661-79.
Benardete, J., 1964, *Infinity*. Oxford, Clarendon Press.
Bennett, J., 1959, "Analytic-Synthetic." *Proceedings of the Aristotelian Society* Vol **59**, pp. 163-188.
——— , 1976, *Linguistic Behavior*. Cambridge, Cambridge University Press.
Bergmann, G., 1953, "Two Cornerstones of Empiricism." *Synthese* Vol **8**, pp. 435-452.
Bertolet, R., 1984a, "Inferences, Names, and Fictions." *Synthese* Vol **58**, pp. 203-218.
——— , 1984b, "Reference, Fiction, and Fictions." *Synthese* Vol **60**, pp. 413-437.
Bigelow, J., and R. Pargetter, 1987, "Beyond the Blank Stare." *Theoria* Vol **53**, pp. 97-114.
Block, N.J., 1981, "Psychologism and Behaviorism." *Philosophical Review* Vol **90**, pp. 5-43.
——— , 1986, "Advertisement for a Semantics for Psychology." *Midwest Studies X: Studies in the Philosophy of Mind* (ed. P. French, T.E. Uehling, and H. Wettstein), Minneapolis, University of Minnesota Press, pp. 615-78.
Boër, S., 1978, "Attributive Names." *Notre Dame Journal of Formal Logic* Vol **19**, pp. 177-85.
——— , 1985, "Substance and Kind: Reflections on the New Theory of Reference." *Analytical Philosophy in Comparative Perspective* (ed. B.-K. Matilal and J. Shaw), Dordrecht, D. Reidel, pp. 103-50.
——— , and W.G. Lycan, 1975a, "Knowing Who." *Philosophical Studies* Vol **28**, pp. 299-344.
——— , 1975b, review of *Counterfactuals*. *Foundations of Language* Vol **13**, pp. 145-151.
——— , 1980, "Who, Me?" *Philosophical Review* Vol **89**, pp. 427-466.
——— , 1986, *Knowing Who*. Cambridge, MA, Bradford Books / MIT Press.
Boghossian, P., 1992, "Analyticity." Unpublished Xerox.
Braun, D., 1993, "Empty Names." *Noûs* Vol **27**, pp. 449-69.

BIBLIOGRAPHY

Burge, T., 1974, "Demonstrative Constructions, Reference, and Truth." *Journal of Philosophy* Vol 71, pp. 205-33.

———, 1975, "On Knowledge and Convention." *Philosophical Review* Vol 84, pp. 249-255.

———, 1979, "Semantical Paradox." *Journal of Philosophy* Vol 76, pp. 169-198. Reprinted in Martin (1984).

Callaway, H., 1980, "Semantic Theory and Language: A Perspective." *Philosophical Topics* Supplementary Volume, pp. 61-70.

———, 1984, "Meaning without Analyticity." *Logique et Analyse* Vol 28, pp. 41-60.

Carnap, R., 1947/1956, *Meaning and Necessity*. Chicago, University of Chicago Press; 2nd edition 1956.

Carney, J.D., 1977, "Fictional Names." *Philosophical Studies* Vol 32, pp. 383-91.

Castañeda, H.-N., 1966, "He: A Study in the Logic of Self-Consciousness." *Ratio* Vol 8, pp. 130-57.

———, 1967, "Indicators and Quasi-Indicators." *American Philosophical Quarterly* Vol 4, pp. 1-16.

———, 1974, "Thinking and the Structure of the World." *Philosophia* Vol 4, 3-40; reprinted in Castañeda (1989).

———, 1979, "Fiction and Reality: Their Basic Connections." *Poetica* Vol 8, pp. 31-62; reprinted in Castañeda (1989).

———, 1989, *Thinking, Language, and Experience*. Minneapolis, University of Minnesota Press.

Chisholm, R.M., 1967, "Identity through Possible Worlds: Some Questions." *Noûs* Vol 1, pp. 1-8. Reprinted in Loux (1979).

Clark, R., 1970, "Concerning the Logic of Predicate Modifiers." *Noûs* Vol 4, pp. 311-335.

Coburn, R., 1986, "Individual Essences and Possible Worlds." French, Uehling and Wettstein (1986), pp. 165-183.

Cohen, L.J., 1985, "A Problem about Ambiguity in Truth-Theoretical Semantics." *Analysis* Vol 45, pp. 129-134.

Conee, E., 1982, "Against Moral Dilemmas." *Philosophical Review* Vol 91, pp. 87-97.

———, 198 [[DD Conference paper]]

Cresswell, M.J., 1972, "The World Is Everything That Is the Case." *Australasian Journal of Philosophy* Vol 50, pp. 1-13; reprinted in Loux (1979).

———, 1973, *Logics and Languages*. London, Methuen.

———, 1974, "Adverbs and Events." *Synthese* Vol **28**, pp. 455-481.
———, 1978, "Semantic Competence." *Meaning and Translation* (ed. F. Guenther and M. Guenther-Reutter), London, Duckworth, pp. 9-43. Reprinted in *Semantical Essays: Possible Worlds and Their Rivals*, Dordrecht, Kluwer Academic Publishers, 1988.
———, 1985, *Structured Meanings*. Cambridge, MA, Bradford Books / MIT Press.
———, 1990, *Entities and Indices*. Dordrecht, Kluwer Academic Publishers.
Crittenden, C., (1991), *Unreality*. Ithaca, Cornell University Press.
Currie, G., 1986, "Fictional Names." *Australasian Journal of Philosophy* Vol **66**, pp. 471-88.
———, 1990, *The Nature of Fiction*. Cambridge, Cambridge University Press.
Davidson, D., 1967a, "The Logical Form of Action Sentences." *The Logic of Decision and Action* (ed. N. Rescher), Pittsburgh, University of Pittsburgh Press, pp. 81-95. Reprinted in Davidson and Harman (1975).
———, 1967b, "Truth and Meaning." *Synthese* Vol **17**, pp. 304-323. Reprinted in Davidson (1984), pp. 17-36.
———, 1968, "On Saying That." *Synthese* Vol **19**, pp. 130-146. Reprinted in Davidson and Harman (1975), and in Davidson (1984).
———, 1984, *Inquiries into Truth and Interpretation*. Oxford, Oxford University Press.
———, 1986, "A Nice Derangement of Epitaphs." *Truth and Interpretation: Perspectives on the Philosophy of Donald Davidson* (ed. E. LePore), Oxford, Basil Blackwell.
———, and G. Harman (eds.), 1972, *Semantics of Natural Language*. Dordrecht, D. Reidel.
———, and G. Harman (eds.), 1975, *The Logic of Grammar*. Encino, CA, Dickenson.
———, and J. Hintikka (eds.), 1969, *Words and Objections: Essays on the Work of W.V. Quine*. Dordrecht, D. Reidel.
Dennett, D.C., 1968, "Geach on Intentional Identity." *Journal of Philosophy* Vol **65**, pp. 335-41.
———, 1978, "Brain Writing and Mind Reading." In *Brainstorms*, Montgomery, VT, Bradford Books, pp. 39-50.

Devitt, M., 1974, "Singular Terms." *Journal of Philosophy* Vol 71, pp. 183-205.
———, 1981, *Designation*, Columbia University Press, New York.
———, 1989, "Against Direct Reference." *Midwest Studies in Philosophy, Vol. XIV: Contemporary Perspectives in the Philosophy of Language II* (ed. P.A. French, T.E. Uehling and H. Wettstein), Minneapolis, University of Minnesota Press, pp. 206-40.
———, 1990, "On Removing Puzzles About Belief Ascription." *Pacific Philosophical Quarterly* Vol 71, pp. 165-81.
———, 1992, "A Critique of the Case for Semantic Holism." Unpublished Xerox.
———, and K. Sterelny, 1987, *Language and Reality: An Introduction to the Philosophy of Language.* Cambridge, MA, Bradford Books / MIT Press.
Donnellan, K., 1970, "Proper Names and Identifying Descriptions." *Synthese* Vol 21, pp. 335-358.
———, 1974, "Speaking of Nothing." *Philosophical Review* Vol 83, pp. 3-31.
Dretske, F., 1988, *Explaining Behavior.* Cambridge, MA, Bradford Books / MIT Press.
Dummett, M., 1973, *Frege: Philosophy of Language.* New York, Harper and Row.
Dunn, M., and N. Belnap, 1968, "The Substitution Interpretation of the Quantifiers." *Noûs* Vol 2, pp. 177-85.
Edelberg, W., 1986, "A New Puzzle About Intentional Identity." *Journal of Philosophical Logic* Vol 15, pp. 1-25.
———, forthcoming, "A Perspectivalist Theory of the Attitudes." *Noûs*.
Ehrman, M. (1966), *The Meanings of the Modals in Present-Day American English.* The Hague, Mouton.
Evans, G., and J. McDowell (eds.), 1976, *Truth and Meaning.* Oxford, Oxford University Press.
Feldman, F., 1971, "Counterparts." *Journal of Philosophy* Vol 68, pp. 406-409.
Field, H., 1978, "Mental Representation." *Erkenntnis* Vol 13, pp. 9-61.
———, 1980, *Science without Numbers.* Princeton, NJ, Princeton University Press.

Fine, K., 1977, "Postscript: Prior on the Construction of Possible Worlds and Instants." In Prior and Fine (1977), pp. 116-161.
———, 1978, "Model Theory for Modal Logic Part II: The Elimination of the De Re." *Journal of Philosophical Logic* Vol 7, pp. 277-306.
———, 1982, "The Problem of Nonexistents, I." *Topoi* Vol 1, pp. 97-140.
———, 1984, Critical Review of Parsons' *Non-Existent Objects*. *Philosophical Studies* Vol 45, pp. 95-142.
Fodor, J.A., 1975, *The Language of Thought*, New York, Crowell.
———, 1978, "Propositional Attitudes." *Monist* Vol 61, pp. 501-23.
———, 1980, "Methodological Solipsism Considered as a Research Strategy in Cognitive Psychology." *Behavioral and Brain Sciences* Vol 3, pp. 63-73.
———, 1987, *Psychosemantics*. Cambridge, MA, Bradford Books / MIT Press.
———, 1990. "Psychosemantics." In Lycan (1990b), pp. 312-27.
Forbes, G., 1985, *The Metaphysics of Modality*. Oxford, Oxford University Press.
———, 1986, "In Defense of Absolute Essentialism." In French, Uehling and Wettstein (1986), pp. 3-31.
———, 1989, *Languages of Possibility: An Essay in Philosophical Logic*. Oxford, Basil Blackwell.
Frege, G., 1891/1952, "Function and Concept." In Geach and Black (1952), pp. 21-41.
———, 1892/1952, "On Sense and Reference." In Geach and Black (1952), pp. 56-78.
———, 1893/1964, *The Basic Laws of Arithmetic: Exposition of the System* (M. Furth, ed. and tr.), Berkeley and Los Angeles, University of California Press.
French, P.A., T.E. Uehling and H. Wettstein (eds.), 1977, *Midwest Studies in Philosophy, Vol. II: Studies in the Philosophy of Language*. Minneapolis, University of Minnesota Press.
———, 1980, *Midwest Studies in Philosophy, Vol. V: Epistemology*. Minneapolis, University of Minnesota Press.
———, 1984, *Midwest Studies in Philosophy, Vol. IX: Causation and Causal Theories*. Minneapolis, University of Minnesota Press.
———, 1986, *Midwest Studies in Philosophy, Vol. XI: Studies in Essentialism*. Minneapolis, University of Minnesota Press.

Geach, P., 1967, "Intentional Identity." *Journal of Philosophy* Vol **64**, pp. 627-32.
———, and M. Black (eds.), 1952, *Translations from the Philosophical Writings of Gottlob Frege*. Oxford, Basil Blackwell.
Gettier, E., (1963), "Is Justified True Belief Knowledge?" *Analysis* Vol **23**, pp. 121-3.
Gibbs, B., (1970), "Real Possibility." *American Philosophical Quarterly* Vol **7**, pp. 340-8.
Gilbert, M., 1981, "Game Theory and *Convention*." *Synthese* Vol **46**, pp. 41-94.
Ginet, C., 1966, "Might We Have No Choice?" *Freedom and Determinism* (ed. K. Lehrer), New York, Random House, pp. 87-104.
Goodman, N., 1968, *Languages of Art*. Indianapolis, Bobbs-Merrill.
Greenbaum, S. (1969), *Studies in English Adverbial Usage*. Coral Gables, FL, University of Miami Press.
Grice, H.P., 1968, "Vacuous Names." In Davidson and Hintikka (1969), pp. 118-45.
———, and P.F. Strawson, 1956, "In Defense of a Dogma." *Philosophical Review* Vol **65**, pp. 141-158.
Haack, S., 1977, "Lewis' Ontological Slum." *Review of Metaphysics* Vol **33**, pp. 415-29.
Hahn, L., and P.A. Schilpp, 1986, *The Philosophy of W.V. Quine*. LaSalle, IL, Open Court.
Hale, S., 1988, "Spacetime and the Abstract/Concrete Distinction." *Philosophical Studies* Vol **53**, pp. 85-102.
Hanson, N.R., 1965, *Patterns of Discovery*. Cambridge, Cambridge University Press.
Harman, G., 1967, "Quine on Meaning and Existence, I." *Review of Metaphysics* Vol **21**, pp. 124-151.
———, 1968a, "Quine on Meaning and Existence, II." *Review of Metaphysics* Vol **21**, pp. 343-68.
———, 1968b, "Three Levels of Meaning." *Journal of Philosophy* Vol **65**, pp. 590-602.
———, 1972a, "Is Modal Logic Logic?" *Philosophia* Vol **2**, pp. 75-84.
———, 1972b, "Logical Form." *Foundations of Language* Vol **9**, pp. 38-65.
———, 1973, *Thought*. Princeton, Princeton University Press.

―――, 1987, *Change in View*. Cambridge, MA, Bradford Books / MIT Press.

Heck, W.C., and W.G. Lycan, 1979, "Frege's Horizontal." *Canadian Journal of Philosophy* Vol **9**, pp. 479-492.

Heidelberger, H., 1980, "Understanding and Truth Conditions." *Midwest Studies in Philosophy V: Studies in Epistemology* (ed. P. French, T.E. Uehling and H. Wettstein), Minneapolis, University of Minnesota Press, pp. 401-10.

Herzberger, H., 1970, "Paradoxes of Grounding in Semantics." *Journal of Philosophy* Vol **67**, pp. 145-67.

Hill, C.S., 1976, "Toward a Theory of Meaning for Belief Sentences." *Philosophical Studies* Vol **30**, pp. 209-26.

Hilpinen, R., 1969, "An Analysis of Relativised Modalities." *Philosophical Logic* (ed. J.W. Davis, D.J. Hockney, and W.K. Wilson), New York, Humanities Press.

Hintikka, K.J.J., 1961, "Modality and Quantification." *Theoria* Vol **27**, pp. 119-28; reprinted in Hintikka (1969a).

―――, 1962, *Knowledge and Belief*. Ithaca, NY, Cornell University Press.

―――, 1969a, *Models for Modalities*. Dordrecht, D. Reidel.

―――, 1969b, "Semantics for Propositional Attitudes." In Hintikka (1969a), pp. 87-111.

―――, 1972, "The Semantics of Modal Notions and the Indeterminacy of Ontology." In Davidson and Harman (1972), pp. 398-414.

―――, 1975a, "Impossible Possible Worlds Vindicated." *Journal of Philosophical Logic* Vol **4**, pp. 475-483; reprinted in *Game-Theoretic Semantics* (ed. E. Saarinen), Dordrecht, D. Reidel, 1979.

―――, 1975, *The Intentions of Intentionality and Other New Models for Modalities*, Dordrecht, D. Reidel.

Howell, R., 1979, "Fictional Objects: How They Are and How They Aren't." *Poetica* Vol **8**, pp. 129-77.

Jamieson, D., 1975, "David Lewis on Convention." *Canadian Journal of Philosophy* Vol **5**, pp. 73-81.

Joos, M. (1964), *The English Verb*. Madison, University of Wisconsin Press.

Kalb, M., and B. Kalb, 1974, *Kissinger*. Boston, Little Brown & Co.

Kapitan, T., 1991, "How Powerful Are We?" *American Philosophical Quarterly* Vol **28**, pp. 331-8.

BIBLIOGRAPHY 313

Kaplan, D., 1964, *Foundations of Intensional Logic.* Unpublished Ph.D. thesis, UCLA.

———, 1973, "Bob and Carol and Ted and Alice." *Approaches to Natural Language* (ed. J. Hintikka, J. Moravcsik and P. Suppes), Dordrecht, D. Reidel, pp. 490-518.

———, 1975, "How to Russell a Frege-Church." *Journal of Philosophy* Vol **72**, pp. 716-729; reprinted in Loux (1979).

———, 1979, "Dthat." In French, Uehling and Wettstein (eds.), *Contemporary Perspectives in the Philosophy of Language*, Minneapolis, University of Minnesota Press, pp. 383-400.

———, 1986, "Opacity." In Hahn and Schilpp (1986), pp. 229-289.

———, 1989, "Afterthoughts." *Themes from Kaplan* (ed. J. Almog, J. Perry and H. Wettstein), Oxford, Oxford University Press, pp. 565-614.

———, 1990, "Words." *Proceedings of the Aristotelian Society* Supplementary Vol **64**, pp. 93-119.

Katz, J.J., 1967, "Some Remarks on Quine on Analyticity." *Journal of Philosophy* Vol **64**, pp. 36-52.

———, 1974, "Where Things Now Stand with the Analytic-Synthetic Distinction." *Synthese* Vol **28**, pp. 283-319.

Kiteley, M., 1981, "Substitution and Reference." *Philosophical Studies* Vol **40**, pp. 221-40.

Kratzer, A., 1977, "What 'Must' and 'Can' Must and Can Mean." *Linguistics and Philosophy* Vol **1**, pp. 113-126.

Kraut, R., 1979, "Attitudes and Their Objects." *Journal of Philosophical Logic* Vol **8**, pp. 197-217.

———, 1987, "Hintikka's Ontology." *Profiles: Jaakko Hintikka* (ed. R.J. Bogdan), Dordrecht, D. Reidel, pp. 261-76.

Kripke, S., 1971, "Identity and Necessity." In Munitz (1971), pp. 135-164.

———, 1972/1980, "Naming and Necessity." In Davidson and Harman (1972), pp. 253-355; republished as *Naming and Necessity*, Cambridge, MA, Harvard University Press, 1980. Page references are to the latter.

———, 1972b, "Vacuous Names and Mythical Kinds." Presented at the Thirteenth Annual Oberlin Colloquium in Philosophy, Oberlin College.

———, 1975, "Outline of a Theory of Truth." *Journal of Philosophy* Vol **72**, pp. 690-716. Reprinted in Martin (1984), pp. 53-81.

―――, 1976, "Is There a Problem about Substitutional Quantification?" In Evans and McDowell (1976), pp. 324-419.
―――, 1979, "A Puzzle About Belief." *Meaning and Use*, (ed. A. Margalit), Dordrecht, D. Reidel, pp. 239-83.
Kuhn, T., 1970, *The Structure of Scientific Revolutions*, Second Edition. Chicago, University of Chicago Press.
Lakoff, G., and M. Johnson, 1980, *Metaphors We Live By*. Chicago, University of Chicago Press.
Lamb, J., 1977, "On a Proof of Incompatibilism." *Philosophical Review* Vol **86**, pp. 20-35.
Lemmon, E.J., 1962, "Moral Dilemmas." *Philosophical Review* Vol **71**, pp. 139-58.
LePore, E., and B. Loewer, 1987, "Dual Aspect Semantics." *New Directions in Semantics*, London, Academic Press, pp. 83-112.
Lewis, D., 1968, "Counterpart Theory and Quantified Modal Logic." *Journal of Philosophy* Vol **65**, pp. 113-26. Reprinted in Loux (1979).
―――, 1969, *Convention*. Cambridge, MA, Harvard University Press.
―――, 1970, "Anselm and Actuality." *Noûs* Vol **4**, pp. 175-88.
―――, 1972, "General Semantics." Davidson and Harman (1972), pp. 169-218.
―――, 1973a, "Causation." *Journal of Philosophy* Vol **70**, pp. 556-67.
―――, 1973b, *Counterfactuals*. Cambridge, MA, Harvard University Press.
―――, 1978, "Truth in Fiction." *American Philosophical Quarterly* Vol **15**, pp. 37-46.
―――, 1979, "Counterfactual Dependence and Time's Arrow." *Noûs* **13**, pp. 455-76.
―――, 1980, "Index, Context and Content." In S. Kanger and S. Öhmann (eds.), *Philosophy and Grammar*. Dordrecht, D. Reidel, pp. 79-100.
―――, 1981, "What Puzzling Pierre Does Not Believe." *Australasian Journal of Philosophy* Vol **59**, pp. 283-9.
―――, 1983, "Individuation by Acquaintance and by Stipulation." *Philosophical Review* Vol **92**, pp. 3-32.
―――, 1986, *On the Plurality of Worlds*. Oxford, Basil Blackwell.
―――, 1990, "Noneism or Allism?" *Mind* Vol **99**, pp. 23-31.
Linsky, B., and E. Zalta, 1991, "Is Lewis a Meinongian?" *Australasian Journal of Philosophy* Vol **69**, pp. 438-53.

Linsky, L., 1969, "Reference, Essentialism, and Modality." *Journal of Philosophy* Vol **66**, pp. 287-300. Reprinted in Linsky (1971); page references are to the latter.
——— (ed.), 1971, *Reference and Modality*. Oxford, Oxford University Press.
Loar, B., 1976, "Two Theories of Meaning." In Evans and McDowell (1976), pp. 138-61.
———, 1981, *Mind and Meaning*. Cambridge University Press, Cambridge.
Lockwood, M., 1971, "Identity and Reference." In Munitz (1971), pp. 199-211.
Loux, M. (ed.), 1979, *The Possible and the Actual*. Ithaca, NY, Cornell University Press.
Lycan, W.G., 1974, "Could Propositions Explain Anything?" *Canadian Journal of Philosophy* Vol 3, pp. 427-435.
———, 1978, "Referential Opacity Explained Away." Talk delivered at the University of Sydney.
———, 1979, "The Trouble with Possible Worlds." In Loux (1979), pp. 274-316.
———, 1980a, "Kripke's Arguments Against the View that Proper Names Abbreviate Descriptions." Unpublished ditto.
———, 1980b, "Thoughts on Stalnaker's Semantics for Belief." Unpublished ditto.
———, 1981, "Form, Function, and Feel." *Journal of Philosophy* Vol **78**, pp. 24-50.
———, 1984, *Logical Form in Natural Language*. Cambridge, MA, Bradford Books / MIT Press.
———, 1987, *Consciousness*. Cambridge, MA, Bradford Books / MIT Press.
———, 1988, *Judgement and Justification*. Cambridge, Cambridge University Press.
———, 1989, "Reply to Baker." *Philosophical Psychology* Vol **2**, pp. 95-100.
———, 1990a, "Mental Content in Linguistic Form." *Philosophical Studies* Vol **58**, pp. 147-154.
——— (ed.), 1990b, *Mind and Cognition: A Reader*. Oxford, Basil Blackwell.

———, 1991, "Homuncular Functionalism Meets PDP." *Philosophy and Connectionist Theory* (ed. W. Ramsey, S.P. Stich and D. Rumelhart), Hillsdale, NJ, Lawrence Erlbaum.
———, 1993, "MPP, RIP." *Philosophical Perspectives*, Vol. 7-8 (ed. J.E. Tomberlin), Atascadero, CA, Ridgeview Publishing.
———, 1994, "Russell's Strange Claim that '*a* Exists' is Meaningless Even When *a* Does Exist." *Russell and Analytic Philosophy* (ed. A. Irvine and G. Wedeking), Toronto, University of Toronto Press.
———, forthcoming, *Real Conditionals*.
———, and R. Nusenoff, 1974, Review of Munitz (1971). *Synthese* Vol **28**, pp. 553-559.
———, and G. Pappas, 1976, "Quine's Materialism." *Philosophia* Vol **6**, pp. 101-130.
———, and S. Shapiro, 1986, "Actuality and Essence." In French, Uehling and Wettstein (1986), pp. 343-377.
Lyons, J., 1963, *Structural Semantics*. Basil Blackwell.
———, 1968, *Introduction to Theoretical Linguistics*. Cambridge, Cambridge University Press.
———, 1977, *Semantics*, 2 vols. Cambridge, Cambridge University Press.
Mackie, J.L. 1977. *Ethics: Inventing Right and Wrong*. New York, Penguin Books.
Mally, E., 1912, *Gegenstandstheoretische Grundlagen der Logik and Logistik*. Leipzig, Barth.
Marcus, R.B., 1960, "Extensionality." *Mind* Vol **69**, pp. 59-62.
———, 1961, "Modalities and Intensional Languages." *Synthese* Vol **13**, pp. 303-322.
———, 1975-1976, "Dispensing with Possibilia." *Proceedings and Addresses of the American Philosophical Association* Vol **44**, pp. 39-51.
———, 1981, "A Proposed Solution to a Puzzle About Belief." *Midwest Studies in Philosophy, Vol. VI: Foundations of Analytic Philosophy* (ed. P. French, T.E. Uehling and H. Wettstein), Minneapolis, University of Minnesota Press, pp. 501-10.
Margolis, J., 1971, *Values and Conduct*. Oxford, Clarendon Press.
Martin, R., ed., 1970, *The Paradox of the Liar*. New Haven, Yale University Press.

———, ed., 1984, *Recent Essays on Truth and the Liar Paradox*. Oxford, Oxford University Press.
McCarthy, T., 1981, "The Idea of a Logical Constant." *Journal of Philosophy* Vol **78**, pp. 499-523.
McCawley, J.D. (1988), *The Syntactic Phenomena of English*, Vol. 1. Chicago, University of Chicago Press.
McConnell, T., 1978, "Moral Dilemmas and Consistency in Ethics." *Canadian Journal of Philosophy* Vol **8**, pp. 269-87.
McGee, V., 1991, *Truth, Vagueness, and Paradox*. Indianapolis, Hackett Publishing.
McGinn, C., 1981, "Modal Reality." *Reduction, Time and Reality* (ed. R. Healey), Cambridge, Cambridge University Press, pp. 143-187.
McKay, T., 1981, "On Proper Names in Belief Ascriptions." *Philosophical Studies* Vol **39**, pp. 287-304.
McKinsey, M., 1978, "Kripke's Objections to Description Theories of Names." *Canadian Journal of Philosophy* Vol **8**, pp. 485-97.
McMichael, A., 1983a, "A Problem for Actualism About Possible Worlds." *Philosophical Review* Vol **92**, pp. 49-66.
———, 1983b, "A New Actualist Modal Semantics." *Journal of Philosophical Logic* Vol **12**, pp 73-99.
———, 1986, "The Epistemology of Essentialist Claims." In French, Uehling and Wettstein (1986), pp. 33-52.
Meinong, A., 1904/1960, "The Theory of Objects." *Realism and the Background of Phenomenology* (ed. R.M. Chisholm), Glencoe, IL, Free Press, pp. 76-117.
Merrill, G.H., 1978, "Formalization, Possible Worlds and the Foundations of Modal Logic." *Erkenntnis* Vol **12**, pp. 305-327.
Miller, R.B., 1989, "Dog Bites Man: A Defence of Modal Realism." *Australasian Journal of Philosophy* Vol **67**, pp. 476-78.
Millikan, R.G., 1984, *Language, Thought, and Other Biological Categories*. Cambridge, MA, Bradford Books / MIT Press.
Montague, R., 1970, "Universal Grammar." *Theoria* Vol **36**, pp. 373-398. Reprinted in Montague (1974).
———, 1968/1974, "Pragmatics." *Contemporary Philosophy: La Philosophie Contemporaire, vol. I* (ed. R. Klibansky), Florence, La Nuova Italia Editrice, pp. 102-22. Reprinted in Montague (1974).
———, 1974, *Formal Philosophy*. New Haven, Yale University Press.

Morgan, J., 1978, "Two Types of Convention in Indirect Speech Acts." In P. Cole (ed.), Syntax and Semantics, Vol. 9: Pragmatics. New York, Academic Press, pp. 261-80.

Morton, A., 1973, "The Possible in the Actual." *Noûs* Vol 7, pp 394-407.

Munitz, M. (ed.), 1971, *Identity and Individuation*. New York, New York University Press.

Naylor, M.B., 1986, "A Note on David Lewis' Realism about Possible Worlds," *Analysis* Vol 46, pp. 28-29.

Nerlich, G., 1991, "How Euclidean Geometry Has Misled Metaphysics." *Journal of Philosophy* Vol 88, pp. 169-189.

Nunberg, G., 1993, "Indexicality and Deixis." *Linguistics and Philosophy* Vol 16, pp. 1-43.

Nute, D., 1976, "David Lewis and the Analysis of Counterfactuals." *Noûs* Vol 10, pp. 353-362.

Palmer, F.R. (1965), *A Linguistic Study of the English Verb*. London, Longmans.

Parsons, T., 1967, "Grades of Essentialism in Quantified Modal Logic." *Noûs* Vol 1, pp. 181-200.

——— , 1969, "Essentialism and Quantified Modal Logic." *Philosophical Review* Vol 78, pp. 35-52. Reprinted in Linsky (1971); page references are to the latter.

——— , 1972, "Some Problems Concerning the Logic of Grammatical Modifiers." In Davidson and Harman (1972).

——— , 1974, "Prolegomenon to a Meinongian Semantics." *Journal of Philosophy* Vol 71, pp. 561-581.

——— , 1980, *Nonexistent Objects*. New Haven, Yale University Press.

Peacocke, C., 1976, "What is a Logical Constant?" *Journal of Philosophy* Vol 73, pp. 221-240.

Perry, J., 1979, "The Problem of the Essential Indexical." *Noûs* Vol 13, pp. 3-21.

——— , 1988, "Cognitive Significance and New Theories of Reference." *Noûs* Vol 22, pp. 1-18.

Plantinga, A., 1974, *The Nature of Necessity*. Oxford, Clarendon Press.

——— , 1978, "The Boethian Compromise." *American Philosophical Quarterly* Vol 15, pp. 129-38.

Pollock, J., "Thinking About An Object." In French, Uehling and Wettstein (1980).

Priest, G., 1987, *In Contradiction*. Dordrecht, Nijhoff.

———, et al. (eds.), *Paraconsistent Logic*. München, Philosophia Verlag.
Prior, A., 1959-60, "Identifiable Individuals." *Review of Metaphysics* Vol **13**, pp. 684-96.
———, and K. Fine, 1977, *Worlds, Times and Selves*. London, Duckworth.
Putnam, H., 1975a, "Dreaming and 'Depth Grammar'." In Putnam (1975c), pp. 304-24.
———, 1975b, "Is Semantics Possible?" In Putnam (1975c), pp. 139-52.
———, 1975c, *Mind, Language and Reality*. Cambridge, Cambridge University Press.
———, 1975d, "The Analytic and the Synthetic." In Putnam (1975c), pp. 33-69.
———, 1975e, "The Meaning of 'Meaning'." In Putnam (1975c), pp. 215-71.
———, 1978, *Meaning and the Moral Sciences*. London, Routledge and Kegan Paul.
Quine, W.V.O., 1948/1963, "On What There Is." *Review of Metaphysics* Vol **2**, pp. 21-38. Reprinted in Quine (1963b); page references are to the latter.
———, 1951/1963, "Two Dogmas of Empiricism." *Philosophical Review* Vol **60**, pp. 20-43. Reprinted in Quine (1963b); page references are to the latter.
———, 1953/1963, "Reference and Modality." In Quine (1963b); page references are to the latter.
———, 1960, *Word and Object*. Cambridge, MA, MIT Press.
———, 1963a, "Carnap and Logical Truth." *The Philosophy of Rudolf Carnap* (ed. P.A. Schilpp), LaSalle, IL, Open Court, pp. 385-406. Reprinted in Quine (1966a).
———, 1963b, *From a Logical Point of View*, Second Edition. New York: Harper Torchbooks.
———, 1966a, *The Ways of Paradox*. New York, Random House.
———, 1966b, "Three Grades of Modal Involvement." In Quine (1966a), pp. 156-74.
———, 1966c, "Truth by Convention." In Quine (1966a), pp. 70-99.
———, 1968, "Ontological Relativity." *Journal of Philosophy* Vol **65**, pp. 185-212. Reprinted in Quine (1969a).
———, 1969a, "Existence and Quantification." In Quine (1969b).

BIBLIOGRAPHY

———, 1969b, *Ontological Relativity and Other Essays*, New York, Columbia University Press.
———, 1969c, "Propositional Objects." In Quine (1969a).
———, 1969d, "Reply to Chomsky." In Davidson and Hintikka (1969), pp. 302-11.
———, 1969e, "Reply to Hintikka." In Davidson and Hintikka (1969), pp. 312-15.
———, 1970, "On the Reasons for Indeterminacy of Translation." *Journal of Philosophy* Vol **67**, pp. 178-183.
———, 1972, "Methodological Reflections on Current Linguistic Theory." In Davidson and Harman (1972), pp. 442-454.
———, 1977, "Intensions Revisited," in French, Uehling and Wettstein (1977), pp. 5-11.
———, 1986a, *Philosophy of Logic*. Second Edition, Cambridge, MA, Harvard University Press.
———, 1986b, "Reply to David Kaplan." In Hahn and Schilpp (1986), pp. 290-4.
———, 1987, "Indeterminacy of Translation Again." *Journal of Philosophy* Vol **84**, pp. 5-10.
Rantala, V., 1975, "Urn Models: A New Kind of Non-Standard Model for First-Order Languages." *Journal of Philosophical Logic* Vol **4**, pp. 455-474.
Rapaport, W.J., 1978, "Meinongian Theories and a Russellian Paradox." *Noûs* Vol **12**, pp. 153-80.
Rescher, N., 1975, *A Theory of Possibility*. Pittsburgh, University of Pittsburgh Press.
———, 1990, *Human Interests*. Stanford, Stanford University Press.
———, and R. Brandom, 1979, *The Logic of Inconsistency*. Totowa, NJ, Rowman and Littlefield.
Richard, M., 1983, "Direct Reference and Ascriptions of Belief." *Journal of Philosophical Logic* Vol **12**, pp. 425-52.
———, 1990, *Propositional Attitudes: An Essay on Thoughts and How We Ascribe Them*. Cambridge, Cambridge University Press.
Richards, T., 1975, "The Worlds of David Lewis." *Australasian Journal of Philosophy* Vol **53**, pp. 105-18.
Rosenberg, J., 1974, *Linguistic Representation*, Dordrecht, D. Reidel.
———, 1978, "Linguistic Roles and Proper Names." *The Philosophy of Wilfrid Sellars: Queries and Examinations* (ed. J. Pitt), Dordrecht, D. Reidel, pp. 189-216.

———, 1981, "On Understanding the Difficulty in Understanding Understanding." *Meaning and Understanding* (ed. H. Parrett and J. Bouveresse), Berlin, Walter de Gruyter, pp. 29-43.
Ross, J., 1987, *The Semantics of Media*. Unpublished doctoral dissertation, Victoria University of Wellington.
Ross, J.F., 1981, *Portraying Analogy*. Cambridge, Cambridge University Press.
Routley, R., 1976, "The Durability of Impossible Objects." *Inquiry* Vol **19**, pp. 247-251.
———, 1980, *Exploring Meinong's Jungle and Beyond*. Canberra, Departmental Monograph #3, Philosophy Department, Research School of Social Sciences, Australian National University.
———, and V. Routley, 1973, "Rehabilitating Meinong's Theory of Objects." *Revue Internationale de Philosophie*, fasc. 2-3, pp. 255-65.
———, et al., 1982a, *Relevant Logics and their Rivals 1*. Atascadero, CA, Ridgeview Publishing.
Russell, B., 1905, Critical Notice of Meinong (ed.), *Untersuchungen zur Gegenstandstheorie und Psychologie*. *Mind* Vol **14**, pp. 530-8.
———, 1905/1956, "On Denoting." *Mind* Vol **14**, pp. 479-493. Reprinted in Russell (1956).
———, 1918-1919/1956, "The Philosophy of Logical Atomism." *The Monist* Vol **28**, pp. 495-527; Vol **29**, pp. 32-63, 190-222, 345-380. Reprinted in Russell (1956), pp. 177-281.
———, 1919, *Introduction to Mathematical Philosophy*. London, Allen and Unwin.
———, 1956, *Logic and Knowledge* (ed. by R.C. Marsh). London, Allen and Unwin.
Saarinen, E., 1978, "Intentional Identity Interpreted: A Case Study of the Relations Among Quantifiers, Pronouns, and Propositional Attitudes." *Linguistics and Philosophy* Vol **2**, pp. 151-223.
Salmon, N., 1986, *Frege's Puzzle*. Cambridge, MA, Bradford Books / MIT Press.
Schiffer, S., 1978, "The Basis of Reference." *Erkenntnis* Vol **13**, pp. 171-206.
Schwayder, D., 1965, *The Stratification of Behavior*. New York, Humanities Press.

Searle, J.R., 1979, "The Logical Status of Fictional Discourse." In *Expression and Meaning*. Cambridge, Cambridge University Press, pp. 58-75.
———, 1983, *Intentionality*. Cambridge, Cambridge University Press.
Sellars, W., 1948, "Concepts as Involving Laws and Inconceivable without Them." *Philosophy of Science* Vol **15**, pp. 287-315; reprinted in *Pure Pragmatics and Possible Worlds: The Early Essays of Wilfrid Sellars* (ed. J. Sicha), Atascadero, CA, Ridgeview Publishing.
———, 1956, "Empiricism and the Philosophy of Mind." *Minnesota Studies in the Philosophy of Science* vol. 1, Minneapolis, University of Minnesota Press, pp. 253-329. Reprinted in Sellars (1963a).
———, 1963a, *Science, Perception, and Reality*. London, Routledge and Kegan Paul.
———, 1963b, "Some Reflections on Language Games." In Sellars (1963a), pp. 321-58.
———, 1967, *Science and Metaphysics*, London, Routledge and Kegan Paul.
———, 1969, "Language as Thought and as Communication." *Philosophy and Phenomenological Research* Vol **29**, pp. 506-27.
———, 1973, "Reply to Quine." *Synthese* Vol **26**, pp. 122-45.
Sharlow, M.F., 1988, "Lewis' Modal Realism: A Reply to Naylor," *Analysis* Vol **48**, pp. 13-15.
Sicha, J., 1974, *A Metaphysics of Elementary Mathematics*. Amherst, MA, University of Massachusetts Press.
Simmons, K., 1993, *Universality and The Liar*. Cambridge, Cambridge University Press.
Skyrms, B., 1981, "Tractarian Nominalism." *Philosophical Studies* Vol **40**, pp. 199-206.
Slote, M., 1982, "Selective Necessity and the Free-Will Problem." *Journal of Philosophy* Vol **79**, pp. 5-24.
Smullyan, A.F., 1948, "Modality and Description." *Journal of Symbolic Logic* Vol **13**, pp. 31-37. Reprinted in Linsky (1971).
Soames, S., 1987, "Direct Reference, Propositional Attitudes, and Semantic Content." *Philosophical Topics* Vol **15**, pp. 47-87.
———, 1988, "Substitutivity." *On Being and Saying: Essays for Richard Cartwright* (ed. J.J. Thomson), Cambridge, MA, MIT Press, pp. 99-132.

BIBLIOGRAPHY

Sober, E., 1982, "Dispositions and Subjunctive Conditionals, or, Why Dormitive Virtues Are No Laughing Matter." *Philosophical Review* Vol **91**, pp. 591-596.

Stalnaker, R., 1968, "A Theory of Conditionals." *Studies in Logical Theory* (ed. Nicholas Rescher), *American Philosophical Quarterly* Monograph Series, No. 2, Oxford, Basil Blackwell, pp. 98-112.

———, 1972, "Pragmatics." Davidson and Harman (1972), pp. 380-397.

———, 1976a, "Possible Worlds." *Noûs* Vol **10**, pp. 65-75; reprinted in Loux (1979).

———, 1976b, "Propositions." *Issues in the Philosophy of Language* (ed. A.F. McKay and D.D. Merrill). New Haven, Yale University Press, pp. 79-91.

———, 1978, "Assertion." *Syntax and Semantics 9: Pragmatics* (ed. P. Cole), New York, Academic Press, pp. 315-32.

———, 1979, "Thoughts." Unpublished Xerox.

———, 1984, *Inquiry*. Cambridge, MA, Bradford Books / MIT Press.

———, 1986, "Counterparts and Identity." In French, Uehling and Wettstein (1986), pp. 121-40.

———, 1987, "Semantics for Belief." *Philosophical Topics* Vol **15**, pp. 177-90.

Stampe, D., 1977, "Towards a Causal Theory of Linguistic Representation." In French, Uehling and Wettstein (1977), pp. 42-63.

———, 1988, 'Need." *Australasian Journal of Philosophy* Vol **66**, pp. 129-60.

Steinberg, D.D. and L. Jakobovits (eds.), 1971, *Semantics*. Cambridge, Cambridge University Press.

Stich, S.P., 1978, "Autonomous Psychology and the Belief-Desire Thesis." *Monist* Vol **61**, pp. 573-591. Reprinted in Lycan (1990b).

Stine, G., 1973, "Essentialism, Possible Worlds, and Propositional Attitudes." *Philosophical Review* Vol **82**, pp. 471-482.

Swinburne, R., 1970, *The Concept of Miracle*. London, Macmillan.

Tarski, A., 1933/1956, "The Concept of Truth in Formalized Languages." In *Logic, Semantics, Metamathematics* (ed. and tr. J.H. Woodger), Oxford, Clarendon Press, pp. 152-197. (Originally published, in Polish, in *Prace Towarzystwa Naukowego Warszawskiego, Wydzial III, No. 34* [1933], pp. vii-116.)

Trigg, R., 1971, "Moral Conflict." *Mind* Vol **80**, pp. 41-55.

Ulm, M., 1978, "Harman's Account of Semantic Paradoxes." *Studies in Language* Vol 2, pp. 379-383.

Unger, P., 1968, "An Analysis of Factual Knowledge." *Journal of Philosophy* Vol. 65, pp. 157-70. Reprinted in *Knowing* (ed. M.D. Roth and L. Galis), New York, Random House.

———, 1984, "Minimizing Arbitrariness: Toward a Metaphysics of Infinitely Many Isolated Concrete Worlds." In French, T.E. Uehling and H. Wettstein (1984), pp. 29-51.

van Fraassen, B., 1973, "Values and the Heart's Command." *Journal of Philosophy* Vol 70, pp. 5-19.

van Inwagen, P., 1975, "The Incompatibility of Free Will and Determinism." *Philosophical Studies* Vol 27, pp. 185-199.

———, 1977, "Creatures of Fiction." *American Philosophical Quarterly* Vol 14, pp. 299-308.

———, 1980, "Indexicality and Actuality." *Philosophical Review* Vol 89, pp. 403-26.

———, 1981, "Why I Don't Understand Substitutional Quantification." *Philosophical Studies* Vol 39, pp. 281-85.

———, 1983, *An Essay on Free Will*. Oxford, Clarendon Press.

———, 1985, "Modal Inference and the Free-Will Problem" unpublished Xerox, Syracuse University.

———, 1986, "Two Concepts of Possible Worlds." French, Uehling and Wettstein (1986), pp. 185-213.

Wallace, J., 1972, "On the Frame of Reference." In Davidson and Harman (1972), pp. 219-252.

———, 1975, "Nonstandard Theories of Truth." In Davidson and Harman (1975), pp. 50-60.

Weinreich, U., 1971, "Explorations in Semantic Theory." In Steinberg and Jakobovits, pp. 308-28.

Wertheimer, R., 1972, *The Significance of Sense*. Ithaca, NY, Cornell University Press.

Wettstein, H., 1991, *Has Semantics Rested on a Mistake? and other Essays*. Stanford, Stanford University Press.

White, A.R., 1975, *Modal Thinking*. Ithaca, NY, Cornell University Press.

Wiggins, D., 1967, *Identity and Spatio-Temporal Continuity*. Oxford, Basil Blackwell.

———, 1971, "On Sentence-Sense, Word-Sense, and Difference of Word Sense: Towards a Philosophical Theory of Dictionaries." In Steinberg and Jakobovits, pp. 14-34.

———, 1973, "Towards a Reasonable Libertarianism." *Essays on Freedom of Action* (ed. T. Honderich), London, Routledge and Kegan Paul, pp. 31-61.

Williams, B., 1973, "Ethical Inconsistency." *Problems of the Self*, Cambridge, Cambridge University Press.

Wilson, N.L., 1959, "Substances Without Substrata." *Review of Metaphysics* Vol 12, pp. 521-39.

Yagisawa, T., 1988, "Beyond Possible Worlds." *Philosophical Studies* Vol 53, pp. 175-204.

———, 1993, "A Semantic Solution to Frege's Puzzle," *Philosophical Perspectives*, Vol. 7 (ed. J.E. Tomberlin), Atascadero, CA, Ridgeview Publishing.

Zalta, E., 1983, *Abstract Objects: An Introduction to Axiomatic Metaphysics*. Dordrecht, D. Reidel.

———, 1988, *Intensional Logic and the Metaphysics of Intentionality*. Cambridge, MA, Bradford Books / MIT Press.

Ziff, P., 1972, *Understanding Understanding*. Ithaca, Cornell University Press.

———, 1977, "About Proper Names." *Mind* Vol 86, pp. 319-32.

Zwicky, A., and J. Sadock, 1975, "Ambiguity Tests and How to Fail Them. *Syntax and Semantics, Vol. 4* (ed. J. Kimball), London, Academic Press.

NAME INDEX

Ackerman, D.F. 137, 142
Adams, R.M. 26, 36, 37, 45, 58, 65, 96, 110, 118, 119, 121-123
Almog, J. 137
Armstrong, D.M. 15, 25, 36, 54-58, 61-67, 84, 97, 98, 99, 120, 219, 256
Aune, B. 174
Austin, D. 140
Baker, L.R. 66, 123, 137, 223
Bar-On, D. 32
Belnap, N. 204, 212
Benacerraf, P. 78, 83, 277
Benardete, J. 84
Bennett, J. 144, 249
Bergmann, G. 249
Bertolet, R. 114
Bigelow, J. 189
Block, N.J. 66, 148, 179, 205, 221, 275
Boër, S. 17, 80, 85, 137, 142, 144, 145, 150, 162, 219
Boghossian, P. 273, 274, 277-279
Brandom, R. 39
Braun, D. 157
Brentano, F. 160, 161
Burge, T. 225, 271
Callaway, H. 171, 249
Carnap, R. 11, 25, 26, 45, 46, 78, 100, 203, 236, 258, 278

Carney, J.D. 120
Castañeda, H.-N. 6, 14, 26, 45, 58, 63, 66, 74, 114, 120, 146
Chisholm, R.M. 76, 97, 98, 118, 121
Clark, R. 115, 116, 118, 120, 141, 153, 239, 246, 247
Coburn, R. 118, 121
Cohen, L.J. 285
Conee, E. 193
Cresswell, M.J. 3, 9, 17, 25, 46, 47, 51, 52, 86, 87, 203-206, 208, 209, 218, 219, 239, 240, 242, 243, 245, 246, 276
Currie, G. 66, 67, 118, 119, 123
Davidson, D. 142, 145, 203, 222, 225-227, 233, 234, 239, 241, 244, 245-247, 285
Dennett, D.C. 126, 144, 223
Descartes, R. 6, 32, 33, 51, 153
Devitt, M. 97, 138, 140, 144, 150, 152, 153, 157, 162, 276-278, 280
Donnellan, K. 135, 157
Dretske, F. 143
Duhem, P. 255
Dummett, M. 140

NAME INDEX

Dunn, M. 204, 212
Edelberg, W. 127-129
Ehrman, M. 172
Elgin, K. 113
Entwisle, R. 89
Feldman, F. 76
Field, H. 15, 63, 96, 129, 143, 148, 257, 290, 296
Fine, K. 14, 37, 38, 45, 63, 87, 103, 114, 121, 244, 272
Fodor, J.A. 143, 144, 146, 148
Forbes, G. 14, 98, 103, 120
Frege, G. 45, 52, 74, 135, 140, 141, 152, 153, 159, 163-165, 203, 207-212, 216, 217, 233, 297, 298, 300
French, P.A. 225, 227
Geach, P.T. 111, 124-129
Gettier, E. 179, 182
Gibbs, B. 174
Gilbert, M. 246, 271
Ginet, C. 186
Gödel, K. 85
Greenbaum, S. 172
Grice, H.P. 249, 250, 294
Haack, S. 80
Hale, S. 15
Hand, M. 35, 219
Hanson, N.R. 258
Harman, G. 9, 10, 55, 143, 144, 209, 236, 242, 246, 250, 252-254, 259, 264, 266-268

Heck, W.C. 207, 210
Heidelberger, H. 204, 223-227
Hill, C.S. 142
Hilpinen, R. 175
Hintikka, K.J.J. 3, 8, 9, 18, 26, 39, 45, 46, 111, 126, 127, 140-142, 203, 233, 285
Hobbes, T. 264
Howell, R. 110, 111, 114, 118
Hume, D. 182
Jamieson, D. 271
Johnson, M. 288
Joos, M. 172
Kapitan, T. 187
Kaplan, D. 36, 104, 110, 114, 135, 137, 152, 203
Katz, J.J. 249
Kiteley, M. 152, 219
Kratzer, A. 174
Kraut, R. 119, 142, 211
Krauschumm 219
Kripke, S. 6, 8, 17, 19, 37, 38, 76-78, 95, 96, 97, 102, 104, 110, 114, 118, 121, 122, 135-137, 139, 140, 142, 145, 146, 149, 152, 157, 171, 204
Kuhn, T. 299
Lakoff, G. 288
Lamb, J. 186
Leibniz, G.W. 3, 64, 65, 233

NAME INDEX

Leibniz, G.W. 3, 64, 65, 233
Lemmon, E.J. 193
LePore, E. 148
Lewis, D.K. 3, 5-7, 9, 15, 16, 18, 19, 25-36, 38-41, 46, 48, 49, 60, 64, 73, 75-91, 96, 98, 99, 111, 112, 116, 118, 119, 123, 125, 148, 158, 171, 183-185, 194, 203, 206, 208, 209, 218, 219, 239, 253, 257, 271-273, 277, 280, 300
Linsky, B. 30
Linsky, L. 16, 100-104
Loar, B. 48, 222, 223
Lockwood, M. 150
Loewer, B. 35, 104, 148
Lyons, J. 172, 285, 288
Mackie, J.L. 15, 111
Mally, E. 26
Mannison, D. 59
Marcus, R.B. 18, 19, 101, 135, 137, 216
Margolis, J. 174, 192, 193
Martin, R. 54
McCarthy, T. 237, 241
McCawley, J.D. 172
McConnell, T. 193
McGinn, C. 46
McKay, T. 137
McKinsey, M. 135
McMichael, A. 104, 112, 113, 117-119, 121, 123, 125, 126, 157

Meinong, A. 4-9, 11, 12, 16, 26, 27, 40, 45, 65, 81, 88, 160, 161
Mill, J.S. 135, 152, 218, 222
Miller, R.B. 89
Millikan, R.G. 143, 144
Montague, R. 8, 9, 203, 233, 239, 285
Moore, G.E. 212, 254
Morgan, J. 174
Morton, A. 113, 123
Naylor, M.B. 39
Nerlich, G. 255
Nunberg, G. 162, 225
Nusenoff, R. 17
Nute, D. 80
Palmer, F.R. 172
Pappas, G. 53
Parsons, T. 14, 26, 27, 39, 45, 46, 58, 63, 74, 103, 114, 158, 239
Peacocke, C. 237, 241, 242
Perry, J. 115, 116, 118-120, 123, 137, 148
Pigden, C. 89
Plantinga, A. 26, 45, 52, 53, 55, 58, 84, 110, 121, 137, 139, 140, 142
Plato 67, 178, 233
Pollock, J. 111
Prior, A. 14, 87, 97

NAME INDEX

Putnam, H.   120, 146, 178, 179, 259, 268, 280, 297
Quine, W.V.   4-10, 13, 17, 38, 46-48, 54, 55, 99, 100-105, 140, 151, 211, 215, 236, 237, 242-244, 249-252, 254, 255, 257-259, 263- 268, 271, 273, 274
Quinn, P.   50, 60
Ramsey, F.P.   118
Rantala, V.   18
Rapaport, W.J.   14
Rescher, N.   6, 17, 37, 39, 46, 76, 77, 79, 83, 96, 110, 178
Richard, M.   17, 25, 37, 137, 140, 150
Richards, T.   28, 59, 77-80, 83, 86
Rosenberg, J.F.   147, 219, 223
Ross, J.   64
Ross, J.F.   285-302
Routley, R.   6, 7, 14, 25-27, 30, 39, 63, 64, 73, 74, 79, 110, 111, 114
Russell, B.   4-6, 16, 17, 26, 52, 61, 88, 97, 99, 112, 135, 142, 149, 156, 165, 203, 233, 234, 254
Sadock, J.   288
Salmon, N.   137, 155, 157
Schiffer, S.   135, 137
Schwayder, D.   271
Searle, J.R.   67, 97-99, 120
Sellars, W.   11, 17, 142, 143, 145, 146, 149, 165, 278
Shapiro, S.   36, 38, 59, 109-111, 118, 119
Sharlow, M.F.   39
Shepard, J.   204
Sicha, J.   17, 19
Skyrms, B.   25, 55, 83, 110
Slote, M.A.   186, 187
Smullyan, A.F.   17, 102
Smyth, R.   113
Soames, S.   137
Sober, E.   244
Stalnaker, R.   3, 9, 18, 26, 30, 35, 46, 53, 60, 80, 89, 142, 144, 203, 206, 208, 209
Stampe, D.   144, 175
Sterelny, K.   144
Stich, S. P.   146, 148
Stine, G.   51
Strawson, P.F.   159, 249, 250
Swinburne, R.   182, 184, 185
Tarski, A.   61, 233
Taschek, W.   273
Taylor, K.   113
Taylor, E.   204, 205
Tichy, P.   98, 111
Trigg, R.   193
Unger, P.   15, 179-182
van Inwagen, P.   15, 30, 33-36, 38, 84, 88, 113, 114, 120, 186-189, 204
Wallace, J.   13, 299

Weinreich, U. 285
Wertheimer, R. 174, 176
Wettstein, H. 137
White, A.R. 115, 116, 118-121, 123, 174, 175, 176, 178, 190, 192-194
Wiggins, D. 97-99, 120, 186, 300
Williams, B. 193
Wilson, N.L. 97, 118
Wittgenstein, L. 25, 56, 64, 65, 222, 233, 278, 283
Yagisawa, T. 39, 153
Zalta, E. 14, 27, 30, 46, 87
Ziff, P. 162, 190, 191, 193, 219
Zwicky, A. 288

SUBJECT INDEX

Accessibility relation 50, 123, Chapter 8 *passim*
Accident 178-182, 187
Actualism 15, Chapters 2-4 *passim*, 99, 116, 118, 119
Analyticity, analytic truth Chapters 11-12 *passim*
Causal-Historical theory of referring 119-121, 127, 128
Combinatorialism 47, 48, 50, 51, 55, 58, 59, 61, 65
Concretism 16, 25, 27, 30, 32, 33, 38, 39, 61, Chapter 4 *passim*, 99, 110, 118
Definition 249, 250, 251, 256, 258-260, Chapter 12 *passim*
Description theories (of referring expressions) 36, 59, 97
Direct reference Chapter 7 *passim*
"Dormitive virtue" explanations 10, 243, 244
Dot quotes 142, 145, 147, 151, 152
Ersatz approach (to possibilia) 15, 16, 19, 25-31, 33-41, Chapter 3 *passim*, 87, 89-91, 104, 112, 122, 158, 159
linguistic ersatzism 39, 65, 87, 96
Essences 36, Chapters 5-6 *passim*, 158
Essentialism Chapters 5-6 *passim*
Existence Chapter 1-2 *passim*, 49, 55, 57, 62, 64, 65, 83, 84, 103, 109, 121, 153, 156, 160, 214, 215, 265, 296, 297, 300, 302
see also Nonexistence
Fictions 61, 63, 65, 114, 179
Figurative language Chapter 13 *passim*
Free will 185-189
Frege's horizontal 207-209, 211, 217
Haecceitism 35-37, 91, 95, 96, 98, 99, 103, 109, 110, 118, 123, 126, 127, 129
for nonexistents Chapter 6 *passim*
Impossibilia 14, 30, 38-40, 64, 87, 88
Indeterminacy of translation 13, 55, 179, 249, 252, 264, 265

"Intentional identity" 124, 127, 128
Kripke's Puzzle Chapter 7 *passim*
Lexical Atomism Chapter 13 *passim*
Logical constants 233, 237, 240-242, 259
Logical form 137, 138, 212, 233, 240, 244, 245, 247, 279, 284, 285, 301
Logical truth 237, 242, 263, 270, 274, 275, 278, 279
Luck 33, 177, 218
Mature Lewis 29, 30, 73, 75, 77, 79, 80, 83, 90, 99, 158
"Meaning postulates" 235-237, 243, 244, 246, 247, 256, 258, 275
Metaphor 284, 286, 291-294
Modal auxiliaries (in natural language) Chapter 8 *passim*
Modal primitives 38, 40, 75, 87-91
Moral "ought" 189-195
"Muteness," see Semantical Muteness
Naturalism 36, 55, 56, 58
Nonexistence Chapters 1-2 *passim*
see also Existence
Nonliteral meaning Chapter 13 *passim*

Occam's Razor 7, 9, 10
Paraphrastic approach (to possibilia) 14-18, 73
counterfactual approach 17, 19
Possibilia Chapters 1-4 *passim*
Possible worlds, see Worlds
Primitive modality, see Modal primitives
Proper names 110, 135, 147, 152, 158, 161
Propositions 11, 15, 17, 32, 36-38, 45, 58-61, 67, 87, 109, 111, 112, 138, 151, 154, 159, 203, 224, 249, 251-254
Quantification 9, 11, 14, 15, 17, 18, 25, 27, 46, 59, 65, 73, 86, 100, 101, 126, 138, 176, 189, 211, 213, 217, 241
Quantifier-Reinterpreting approach (to possibilia) 15, 16, 18, 19, 25, 27, 45, 73, 74
Relentlessly Meinongian approach (to possibilia) 73, 75, 76, 78, 83, 90
Representational theory of thoughts 142-149
Rule of Effability 137-139

SUBJECT INDEX

Semantical muteness 211-228
Substitutional quantification 18, 211
Synonymy 149, 206, 236, 250-252, 258, 263-265
Transworld identity 32, 35-37, 59, 76, 91, 103, 109, 111, 112, 117-119, 125
Truth-conditional semantics 3, 19, 138, 147, 153, 155, 156, 161-164, Chapters 9-10 *passim*, 273-276, 299-302
  and indexicals 223-228
Worlds
  possible 3, 5, 8, 9, 14-19, 25, 27, 30, 31, 35, 37, 39, 45, 47, 48, 50, 51, 55, 58, 61-64, 66, 67, 78-80, 83, 86, 87, 96, 103, 104, 109, 113, 114, 126, 135, 138, 171, 203-206, 208, 209, 218, 221, 222, 227, 249, 253-255
  impossible 38, 39, 41, 59, 75, 89, 91

Studies in Linguistics and Philosophy

1. H. Hiż (ed.): *Questions*. 1978   ISBN 90-277-0813-4; Pb: 90-277-1035-X
2. W. S. Cooper: *Foundations of Logico-Linguistics*. A Unified Theory of Information, Language, and Logic. 1978
      ISBN 90-277-0864-9; Pb: 90-277-0876-2
3. A. Margalit (ed.): *Meaning and Use*. 1979   ISBN 90-277-0888-6
4. F. Guenthner and S.J. Schmidt (eds.): *Formal Semantics and Pragmatics for Natural Languages*. 1979   ISBN 90-277-0778-2; Pb: 90-277-0930-0
5. E. Saarinen (ed.): *Game-Theoretical Semantics*. Essays on Semantics by Hintikka, Carlson, Peacocke, Rantala, and Saarinen. 1979   ISBN 90-277-0918-1
6. F.J. Pelletier (ed.): *Mass Terms: Some Philosophical Problems*. 1979
      ISBN 90-277-0931-9
7. D. R. Dowty: *Word Meaning and Montague Grammar*. The Semantics of Verbs and Times in Generative Semantics and in Montague's PTQ. 1979
      ISBN 90-277-1008-2; Pb: 90-277-1009-0
8. A. F. Freed: *The Semantics of English Aspectual Complementation*. 1979
      ISBN 90-277-1010-4; Pb: 90-277-1011-2
9. J. McCloskey: *Transformational Syntax and Model Theoretic Semantics*. A Case Study in Modern Irish. 1979   ISBN 90-277-1025-2; Pb: 90-277-1026-0
10. J. R. Searle, F. Kiefer and M. Bierwisch (eds.): *Speech Act Theory and Pragmatics*. 1980   ISBN 90-277-1043-0; Pb: 90-277-1045-7
11. D. R. Dowty, R. E. Wall and S. Peters: *Introduction to Montague Semantics*. 1981; 5th printing 1987   ISBN 90-277-1141-0; Pb: 90-277-1142-9
12. F. Heny (ed.): *Ambiguities in Intensional Contexts*. 1981
      ISBN 90-277-1167-4; Pb: 90-277-1168-2
13. W. Klein and W. Levelt (eds.): *Crossing the Boundaries in Linguistics*. Studies Presented to Manfred Bierwisch. 1981   ISBN 90-277-1259-X
14. Z. S. Harris: *Papers on Syntax*. Edited by H. Hiż. 1981
      ISBN 90-277-1266-0; Pb: 90-277-1267-0
15. P. Jacobson and G. K. Pullum (eds.): *The Nature of Syntactic Representation*. 1982   ISBN 90-277-1289-1; Pb: 90-277-1290-5
16. S. Peters and E. Saarinen (eds.): *Processes, Beliefs, and Questions*. Essays on Formal Semantics of Natural Language and Natural Language Processing. 1982
      ISBN 90-277-1314-6
17. L. Carlson: *Dialogue Games*. An Approach to Discourse Analysis. 1983; 2nd printing 1985   ISBN 90-277-1455-X; Pb: 90-277-1951-9
18. L. Vaina and J. Hintikka (eds.): *Cognitive Constraints on Communication*. Representation and Processes. 1984; 2nd printing 1985
      ISBN 90-277-1456-8; Pb: 90-277-1949-7
19. F. Heny and B. Richards (eds.): *Linguistic Categories: Auxiliaries and Related Puzzles*. Volume I: Categories. 1983   ISBN 90-277-1478-9

Volumes 1–26 formerly published under the Series Title: *Synthese Language Library*.

Studies in Linguistics and Philosophy

20. F. Heny and B. Richards (eds.): *Linguistic Categories: Auxiliaries and Related Puzzles.* Volume II: The Scope, Order, and Distribution of English Auxiliary Verbs. 1983 ISBN 90-277-1479-7
21. R. Cooper: *Quantification and Syntactic Theory.* 1983 ISBN 90-277-1484-3
22. J. Hintikka (in collaboration with J. Kulas): *The Game of Language.* Studies in Game-Theoretical Semantics and Its Applications. 1983; 2nd printing 1985
 ISBN 90-277-1687-0; Pb: 90-277-1950-0
23. E. L. Keenan and L. M. Faltz: *Boolean Semantics for Natural Language.* 1985
 ISBN 90-277-1768-0; Pb: 90-277-1842-3
24. V. Raskin: *Semantic Mechanisms of Humor.* 1985
 ISBN 90-277-1821-0; Pb: 90-277-1891-1
25. G. T. Stump: *The Semantic Variability of Absolute Constructions.* 1985
 ISBN 90-277-1895-4; Pb: 90-277-1896-2
26. J. Hintikka and J. Kulas: *Anaphora and Definite Descriptions.* Two Applications of Game-Theoretical Semantics. 1985 ISBN 90-277-2055-X; Pb: 90-277-2056-8
27. E. Engdahl: *Constituent Questions.* The Syntax and Semantics of Questions with Special Reference to Swedish. 1986 ISBN 90-277-1954-3; Pb: 90-277-1955-1
28. M. J. Cresswell: *Adverbial Modification.* Interval Semantics and Its Rivals. 1985
 ISBN 90-277-2059-2; Pb: 90-277-2060-6
29. J. van Benthem: *Essays in Logical Semantics* 1986
 ISBN 90-277-2091-6; Pb: 90-277-2092-4
30. B. H. Partee, A. ter Meulen and R. E. Wall: *Mathematical Methods in Linguistics.* 1990; Corrected second printing of the first edition 1993
 ISBN 90-277-2244-7; Pb: 90-277-2245-5
31. P. Gärdenfors (ed.): *Generalized Quantifiers.* Linguistic and Logical Approaches. 1987 ISBN 1-55608-017-4
32. R. T. Oehrle, E. Bach and D. Wheeler (eds.): *Categorial Grammars and Natural Language Structures.* 1988 ISBN 1-55608-030-1; Pb: 1-55608-031-X
33. W. J. Savitch, E. Bach, W. Marsh and G. Safran-Naveh (eds.): *The Formal Complexity of Natural Language.* 1987 ISBN 1-55608-046-8; Pb: 1-55608-047-6
34. J. E. Fenstad, P.-K. Halvorsen, T. Langholm and J. van Benthem: *Situations, Language and Logic.* 1987 ISBN 1-55608-048-4; Pb: 1-55608-049-2
35. U. Reyle and C. Rohrer (eds.): *Natural Language Parsing and Linguistic Theories.* 1988 ISBN 1-55608-055-7; Pb: 1-55608-056-5
36. M. J. Cresswell: *Semantical Essays.* Possible Worlds and Their Rivals. 1988
 ISBN 1-55608-061-1
37. T. Nishigauchi: *Quantification in the Theory of Grammar.* 1990
 ISBN 0-7923-0643-0; Pb: 0-7923-0644-9
38. G. Chierchia, B.H. Partee and R. Turner (eds.): *Properties, Types and Meaning.* Volume I: Foundational Issues. 1989 ISBN 1-55608-067-0; Pb: 1-55608-068-9
39. G. Chierchia, B.H. Partee and R. Turner (eds.): *Properties, Types and Meaning.* Volume II: Semantic Issues. 1989 ISBN 1-55608-069-7; Pb: 1-55608-070-0
 Set ISBN (Vol. I + II) 1-55608-088-3; Pb: 1-55608-089-1

Studies in Linguistics and Philosophy

40. C.T.J. Huang and R. May (eds.): *Logical Structure and Linguistic Structure*. Cross-Linguistic Perspectives. 1991 ISBN 0-7923-0914-6; Pb: 0-7923-1636-3
41. M.J. Cresswell: *Entities and Indices*. 1990
    ISBN 0-7923-0966-9; Pb: 0-7923-0967-7
42. H. Kamp and U. Reyle: *From Discourse to Logic*. Introduction to Modeltheoretic Semantics of Natural Language, Formal Logic and Discourse Representation Theory. 1993 ISBN 0-7923-2403-X; Student edition: 0-7923-1028-4
43. C.S. Smith: *The Parameter of Aspect*. 1991
    ISBN 0-7923-1136-1; Pb 0-7923-2496-X
44. R.C. Berwick (ed.): *Principle-Based Parsing*. Computation and Psycholinguistics. 1991 ISBN 0-7923-1173-6; Pb: 0-7923-1637-1
45. F. Landman: *Structures for Semantics*. 1991
    ISBN 0-7923-1239-2; Pb: 0-7923-1240-6
46. M. Siderits: *Indian Philosophy of Language*. 1991 ISBN 0-7923-1262-7
47. C. Jones: *Purpose Clauses*. 1991 ISBN 0-7923-1400-X
48. R.K. Larson, S. Iatridou, U. Lahiri and J. Higginbotham (eds.): *Control and Grammar*. 1992 ISBN 0-7923-1692-4
49. J. Pustejovsky (ed.): *Semantics and the Lexicon*. 1993
    ISBN 0-7923-1963-X; Pb: 0-7923-2386-6
50. N. Asher: *Reference to Abstract Objects in Discourse*. 1993 ISBN 0-7923-2242-8
51. A. Zucchi: *The Language of Propositions and Events*. Issues in the Syntax and the Semantics of Nominalization. 1993 ISBN 0-7923-2437-4
52. C.L. Tenny: *Aspectual Roles and the Syntax-Semantics Interface*. 1994
    ISBN 0-7923-2863-9; Pb: 0-7923-2907-4
53. W.G. Lycan: *Modality and Meaning*. 1994
    ISBN 0-7923-3006-4; Pb: 0-7923-3007-2

Further information about our publications on *Linguistics* are available on request.

Kluwer Academic Publishers – Dordrecht / Boston / London

Q